# INTERNATIONAL CENTRE FOR MECHANICAL SCIENCES

COURSES AND LECTURES - No. 40

LUIGI G. NAPOLITANO
INSTITUTE OF AERODYNAMICS
UNIVERSITY OF NAPOLI, ITALY

AND

OLEG M. BELOTSERKOVSKII
COMPUTING CENTER
ACADEMY OF SCIENCES
MOSCOW, USSR

# COMPUTATIONAL GASDYNAMICS

SPRINGER-VERLAG WIEN GMBH

Originally published by Springer-Verlag Wien-New York in 1975

ISBN 978-3-211-81428-4 ISBN 978-3-7091-2732-2 (eBook)
DOI 10.1007/978-3-7091-2732-2

# CONTENTS

**L.G. NAPOLITANO**

*Institute of Aerodynamics – University of Napoli*
*Italy*

# BASIC EQUATIONS OF GASDYNAMICS

## INTRODUCTION

The evolution of elementary fluid particles (in the 'continuum' sense) is characterized by exchanges of energy, momentum and mass with their environment. When they have comparatively large total energy density a number of "internal rate processes" can also take place. Any elementary fluid particle is in this case a "composite system" whose subsystems interact through a number of exchanges of energy and mass. The mass exchanges derive from alterations of the molecular structures of the constituent gasses, i.e. from chemical reactions. The energy exchanges derive from redistributions of energy among the different degrees of freedom of the molecules, including the "excitation" of internal degrees which are usually "unexcited" at lower energy density levels.

To analize in detail any flow field one must be able to: a) define, and consequently characterize and describe, the "state" of the evolving elementary fluid particle; b) define, characterize and describe the different processes which occur during its evolution (i.e. both the "internal ones" taking place among its subsystems and the "external ones" involving the interaction with the environment); c) formulate the "constraints" under which the evolution takes place (i.e. impose the basic conservation principles and express the balance equations).

Two different approaches can be used: the phenomenological "macroscopic" approach and the "statistical" approach. Each has its own advantages and shortcomings. Phenomenological theories are able to cope with more general situations and lead to more generally applicable results than statistical theories, whose inherent complexity restricts the classes of problems that can be studied. On the other hand the statistical approach leads to quantitave results expressed in terms of physical characteristics of the atoms and molecules constituting the medium, a thing that a phenomenological theory is unable to do.

Hence, in principle, to formulate the entire closed set of basic equations one must use a suitable combination of phenomenological and statistical theories. A possible way of going about it is to start with complex phenomenologies. As a matter of fact, it is the very same macroscopic approach which will make it possible to <u>recognise</u> how complex a phenomenology is, i.e. how many different processes can indeed take place. When this approach has shown the nature, number and types of information which are needed to study in detail the evolution of specific systems, one reverts to statistical theories to obtain them, in explicit form, in terms of the physical properties of the elementary particles constituting the system. The status of the art of statistical theories is not always and consistently up to this task. When such is the case, the still needed information must be obtained from experimental data.

In the present lectures, the macroscopic approach will be dealt with at length, in order to clearly identify and describe the phenomenologies involved. The macroscopic theory will be integrated by results derived from statistical theories for some simple "systems" of significant and relevant interest. No detailed treatment of these theories will however be given, due to obvious time limitations.

Whatever the approach taken (i.e. whether macroscopic or statistical) the present status of the art is such that, with the possible exception of particular and rather simple cases, one cannot avoid making a rather drastic simplifying assumption which, in want of a better terminology, we shall refer to as the "shifting equilibrium" hypothesis. The thorough and detailed discussion of what such an hypothesis really means, of what are its foreseenable limits of validity and of when and how these limits can be formulated "quantitatively" may well form the object of a short course in itself. By limiting ourselves to the macroscopic point of view, we may say that there are particular "conditions" of any given system, called "equilibrium" states which are characterized, on the macroscopic scale, by the absence of any extensive quantity. Such states can be described in terms of a finite number of parameters and have a number of important properties whose study forms the object of "Equilibrium Thermodynamics". By definition, any evolution of the system (simple of composite) involves interaction (i.e. exchanges) with its environment and/or "internal" productions and/or (if it is a composite system) interactions among its subsystems. Hence, in principle, the system is not in "equilibrium" conditions during any of its evolutions. It may however occur that, depending on the "size" of the system and the nature of the evolution, its "states" and their properties may still be described by using "Equilibrium Thermodynamics" results. This in essence, is the "shifting equilibrium" hypothesis. Loosely speaking, it may be said that such hypothesis is valid when the characteristic times for the "equilibration" of a system (resp. subsystem) are much smaller than those associated with its interaction with the surroundings (resp. with other subsystems). Since the former ones are monotonically increasing functions of the "size" of the system, they can always be made very small by reducing it (hence the necessity of the so-called "local" formulation of thermodynamics) and the ultimate limits of applicability of the shifting-equilibrium hypothesis are reached when, for given processes, this size must be so small as to become of the order of the molecular free path and thus invalidate the continuum approach.

The shifting-equilibrium hypothesis uncouples, so to speak, the study of the evolution of a system by allowing it to be made in two seperate stages. First one studies its "equilibrium" properties; then, by assuming that they still apply during the processes, one studies and analizes these processes, a task which falls within the realm of so called "Irreversible (or non-Equilibrium) Thermodynamics". In particular, when applied to composite systems, the shifting equilibrium hypothesis makes it possible to give a thermodynamic description of the system (and of its properties) even when its subsystems interact by exchanging mass and energy.

The general plan of these lectures should now be apparent.

We shall first deal with the 'Equilibrium Thermodynamics" of composite systems among whose subsystems exchanges of mass (i.e. chemical reactions) and of energy (e.g. excitation of internal degrees of freedom) may take place. Upon the shifting-equilibrium hypothesis the "point" characterizing the state of the system will be on a well defined hypersurface of an appropriate "thermodynamic space", whose general properties will be described. Such a composite system may find itself in several possible different "equilibrium conditions" according to the number and types of "internal" processes (i.e. exchanges among its subsystems) which are either inhibited (i.e. "frozen") or "equilibrated". Each such condition corresponds to well defined loci of points on the above mentioned hypersurface, whose relevant properties will be discussed. Other properties of the subject system relevant to its evolution, such as its thermodynamic stability, and some classes of thermodynamic potentials will subsequently be analyzed.

As already mentioned, macroscopic theories fall short of furnishing explicit forms for the functional relationships that they introduce (e.g. the equations of state). Assigning these explicit forms amounts to prescribing what we call a "thermodynamic model" of the system. Such models can be constructed with the help of statistical thermodynamics and/or suitable assumptions suggested experimental evidence. We shall limit ourselves to the presentation of special and simple models which find extended use.

When one knows how to characterize a thermodynamic "state" of a system one knows what is the set of field unknowns during its evolution, provided one "decides" how detailed will have to be the characterization of its "kinematical" state. We shall assume that one mass velocity suffices, i.e. we shall deal with the so-called single-fluid theory. In this theory the mass velocities of each constituent of the mixture is not considered as a "primitive" unknown but is treated, via the diffusion fluxes, as an unknown to be characterized phenomenologically.

For each extensive "primitive" unknown one can formulate a balance equation. The global, differential and "jump" forms of the balance equations shall be developed for an arbitrary extensive quantity first and then specialized to the basic set of field unknowns. The imposition of the basic principles of total mass, momentum (we suppose negligible the gravitational effects) moment of momentum and total energy will subsequently lead to the basic set of field equations. This set of equations is not "closed' as it contains a number of fluxes and production which will have to be expressed, phenomenologically, in terms of the basic set of unknowns.

This will be done to within the limits of linear irreversible thermodynamics whose basic postulates (Curie and Onsager) will be formulated and illustrated.

Before proceeding with the plan outlined above it is appropriate to say a few words about the general philosophy which underlies the present lectures.

The phenomenology involved in continuum flows of gas mixtures may become very complex and, obviously, the actual solution of specific problems can only be done by steps. One starts with simple (even oversimplified) ideal schemes, assesses by comparison with experimental data the limits of applicability of the simpler results and tries subsequently to formulate gradually more complex and adequate schemes. An indispensable prerequisite to all this, however, is the availability of basic equations derived in a manner as general and as inclusive of all possible foreseenable effects as possible in order to be able to both realize the true meaning and implication of an assumed simplified scheme and to known in which direction one should go to try to improve on it.

Generality and inclusiveness were thus taken as guiding goals in the preparation of these notes. This decision was further motivated by the belief that among the scopes of the CISM courses there is also that of stimulating the student's interest in carrying on original researches to further the progress in a given field of fluid-dynamics.

Obviously there is a limit to everything. Thus we have had to make some choices and leave out classes of phenomena. The two more important ones are:phase changes and radiation effects. To partially compensate for this we have included in these notes a number of original, still unpublished results.

## 1. THERMODYNAMIC DESCRIPTION

If capillarity, gravitational and electromagnetic effects are negligible, the state of a mixture of $(n)$ gasses is, to within the limits of validity of the shifting equilibrium hypothesis, sufficiently and uniquely characterized when one prescribes the values of the volume $(V)$ occupied by the mixture, the $(n)$ masses $M_i$ of its constituents (*) and the $(k+1)$ entropies $S_j$ of all of its subsystems which are not in mutual equilibrium with respect to exchanges (**). Equivalently, one can give the total mass $M = \sum_{i=1}^{n} M_i$, $(n-1)$ masses $M_i$, the total entropy $S = \sum_{j=1}^{k+1} S_j$ and the $k$ remaining entropies $S_j$.

The number $(d+1) = (n+k+2)$ shall be referred to as the number of extensive thermodynamic degrees of freedom of the systems whereas $d = n+k+1$ shall be referred to as the

---

(*) Equivalently one can describe the composition of the mixture by means of the number of moles of each constituent.

(**) In a more precise notation one should label these entropies with two indexes, one referring to the type of mass of the sub-system, the other to the particular type of ensemble of degrees of freedom being considered. In a more general formulation (which may be relevant when dealing with multiphase systems) also translation-rotational degrees of freedom of the different species may not be in mutual equilibrium with respect to energy exchanges.

Number of specific degrees of freedom of the system. In the present contest, specific values will always mean quantities referred to the total mass of the mixture.

The actual value $d$ depends obviously on the specific case considered. In particular, for simple gasses $(n=1)$ in thermal equilibrium it is $d=3$.

Upon a basic postulate, the thermodynamic behaviour of the subject composite systems is completely characterized by the energy fundamental relation (*)

$$U = U(S, V, M, M_i, S_j) \tag{1.1}$$

where $U$, the internal energy, is an extensive quantity. The set of extensive parameters

$$[S, V, M, M_i, S_j] \qquad (1 \leq i \leq n-1 \; ; \; 1 \leq j \leq k) \tag{1.2}$$

will be referred to as the natural set of extensive thermodynamic variables for the internal energy.

Denote with a low case letter quantities referred to the unit mass of the mixture (e.g. $s = S/M$ and with $c_i = M_i / M$

(*) A completely parallel development can be made in terms of an entropy fundamental relation. There are situations when this is preferable, however time limitations prevent us from presenting the full dual development.

the mass concentration of the $\iota$-th constituent. The set

$$\left[s, v, c_i, s_j\right] \tag{1.3}$$

will be referred to as the set of natural specific variables of the internal energy.

The function (1.1) is single valued, continuous together with at least its first three derivatives (we shall not consider here phase changes), first order homogeneous, monotonically increasing with $S$ and $M$ and monotonically decreasing with $V$. Upon the well known Euler theorems on homogeneous functions it then follows that:

$$U = M u(s, v, c_i, s_j)$$

$$u = Ts - pv + g_n + \sum_{i=1}^{n-1} (g_i - g_n) c_i + \sum_{j=1}^{k} (\Theta_j - T) s_j \tag{1.4a}$$

$$0 = s dT - v dp + dg_n + \tag{1.4b}$$

$$+ \sum_{i=1}^{n-1} c_i \, d(g_i - g_n) + \sum_{j=1}^{k} s_j \, d(\Theta_j - T)$$

where the thermodynamic absolute temperatures $T$, $\Theta_j$ the thermodynamic pressure $p$ and the electrochemical potentials are defined by:

$$T = \frac{\partial U}{\partial S} = \frac{\partial u}{\partial s} > 0$$

$$-p = \frac{\partial U}{\partial V} = \frac{\partial u}{\partial v} < 0 \tag{1.5a}$$

$$g_n = \frac{\partial U}{\partial M} > 0 \; ; \quad (g_i - g_n) = \frac{\partial U}{\partial M_i} = \frac{\partial u}{\partial c_i}$$

(1.5b)

$$(\theta_j - T) = \frac{\partial U}{\partial S_j} = \frac{\partial u}{\partial s_j}$$

so that

$$dU = TdS - pdV + g_n dM + \sum_{i=1}^{n-1} (g_i - g_n) dM_i + \\ + \sum_{j=1}^{k} (\theta_j - T) dS_j$$

(1.6)

$$du = Tds - pdv + \sum_{i=1}^{n-1} (g_i - g_n) dc_i + \sum_{j=1}^{k} (\theta_j - T) ds_j$$

Equations (1.4a), (1.4b) and (1.6) are known as Euler, Gibbs-Duhem and Gibbs relations respectively.

The quantities defined by eqs. (1.5) are zeroth order homogeneous functions of the natural set (1.2) (i.e. they are intensive quantities) and only "d" of them are independent. The set:

(1.7) $$\left[T, -p, (g_i - g_n), (\theta_j - T)\right]$$

shall be referred to as the set of intensive parameters conjugate to the set (1.3) with respect to the internal energy.

A specific variable and its intensive conjugate quantity constitute a conjugate couple.

The functional dependences of the intensive quantities on the natural specific variables (1.3) yields a set of independent equations of state of the mixture. Since they represent the set of first order derivatives of the energy fundamental relation it follows that the thermodynamic behaviour of the mixture is completely characterized (unless an inessential constant) when $d = (n+k+1)$ independent equations of state are given. Thus: a necessary and sufficient condition for the complete description of a system is the knowledge of a number of independent equations of state equal to the number "$d$" of the specific degrees of freedom of the system. Assigning an explicit form to the fundamental relation (or, equivalently, to "$d$" independent equations of state) amounts to give a particular 'thermodynamic' model for the medium. Some models will be presented and discussed later on, after completing the analysis of the properties of the energy function.

Example 1

The set of energy specific natural variables for the "simple gas" is:

$$s, v$$

The energy fundamental equation is:

$$u = u(s, v)$$

The set of intensive conjugate parameters is the set:

$$T, \ -p$$

The conjugate couples are:

$$(T, s) \ ; \ (-p, v)$$

The Euler, Gibbs and Gibbs-Duhem relations read:

$$u = Ts - pv + g$$

$$du = Tds - pdv$$

$$0 = sdT - vdp + dg$$

In order to simplify the notations we shall often denote by $x_r$ any of the specific variables $(c_i, s_j)$ and by $y_r$ its conjugate parameter so that the fundamental relation will be written as:

$$u = u(s, v, x_r)$$

[with the dummy index running from 1 to $(d-2)$ if "$d$" is the number of specific degrees of freedom of the system] and, by definition:

$$y_r = \frac{\partial u}{\partial x_r} = y_r(x_r, s, v), \quad (\forall r)$$

## 2. EQUILIBRIUM CONDITIONS

From a general point of view, thermodynamic equilibrium exists when on a microscopic level, all fluxes and productions (which shall be defined more rigorously later) of extensive quantities vanish identically within the system and on its boundaries.

A composite system will be in a condition of equilibrium if and only if:

a) each subsystem is itself in equilibrium; b) no exchange of either mass or energy occurs among the subsystems (the exchange of any other extensive quantity, such as the entropy, is necessarily related to that of either energy or mass or both). These exchanges may be prevented by constraints. A constraint may be either active or not: it is active when its elimination breaks the equilibrium of a composite system. An equilibrium will be called "constrained" if the constraints are active.

Let $x_i^{(e)}$ be the values of the extensive variables characterizing a particular equilibrium state $\sigma_e$ of a composite system. We shall call "virtual displacement" $\delta x_i$ any infinitesimal variation of $x_i$ around $x_i^{(e)}$ which is compatible with the constraints.

The following basic postulate applies in the subject case:

"A state of stable equilibrium $\sigma_e$ of an isolated composite system is characterized by the minimum of its internal energy with respect to all virtual displacements from $\sigma_e$. For a state of constrained equilibrium the minimum is a relative minimum which decreases as the number of active constraints decreases. For a state of unconstrained equilibrium the minimum is an a absolute one".

The isolation condition implies that the virtual displacements are such that $\delta S, \delta v, \delta M \equiv 0$. In the subject system the other possible constraints indicate whether or not a chemical reaction (mass exchange process) (*) or excitation of a given degree of freedom of a constituent molecule (energy exchange process) may occur in the mixture. When a given energy exchange is inhibited by a constraint the corresponding virtual displacements $\delta S_i$ are zero and one says that the particular process is "frozen". Similarly, when the constituent gases are inert, all virtual displacements $\delta c_i$ are zero and one says that the reactions are "frozen". When chemical reactions may occur, the $\delta c_i'$ are not all independent as they are subjected to a number of constraints.

(+) Notice that, in the thermodynamic framework, mass diffusion is to be taken as an exchange of mass between the composite system and its surroundings and that, upon the shifting equilibrium hypothesis, the considerations which are developed in the absence of such interaction with the surroundings will also apply when this interaction does take place. Similar remarks apply for the energy diffusion processes.

To formulate explicitly these constraints we shall symbolically indicate a general chemical reaction $(\ell)$ as:

$$\sum_{i=1}^{n} \bar{\nu}_{\ell i}[c_i] \rightleftharpoons \sum_{i=1}^{n} \bar{\nu}'_{\ell i}[c_i] \, ; \qquad (s \leq \ell \leq r)$$

or,

$$\sum_{i=1}^{n} \left[\bar{\nu}'_{\ell i} - \bar{\nu}_{\ell i}\right]\left[c_i\right] = 0$$

where:

$[c_i]$ stands for the element participating in the reaction

$\bar{\nu}_{\ell i}$ is the stoichiometric coefficient of the $i$-th reaction

$\bar{\nu}'_{\ell i}$ is the stoichiometric coefficient of the $i$-th product

The choice of the "sense" of the reaction is quite arbitrary and $\bar{\nu}_{\ell i}$ (resp. $\bar{\nu}'_{\ell i}$) will be zero if the $i$-th element is not included among the reactants (resp. reaction products) of the $\ell$ -th reaction.

If $M_i$ is the molar mass of the $i$-th species, the mass conservation during a reaction implies that :

$$\sum_{i=1}^{n} (\bar{\nu}'_{\ell i} - \bar{\nu}_{\ell i}) M_i = 0$$

By introducing a reference molar mass $M_r$ one defines new quantities $\nu_{\ell i}$ as:

(2.2) $$\nu_{\ell i} = (\bar{\nu}'_{\ell i} - \bar{\nu}_{\ell i}) \frac{M_i}{M_r}$$

which measure, nondimensionally, the stoichiometric mass change of the $i$-th element due to the $\ell$-th reaction. The mass conservation then reads:

$$\sum_{i=1}^{n} \nu_{\ell i} = 0 \; ; \quad (\forall \ell)$$

and there is one such relation for each reaction.

To properly account for the actual chemical kinetic mechanism of certain overall reactions one often needs to formulate a number of additional reactions, describing the 'intermediate steps" through which the process goes. It may happen that not all such reactions are independent [in a system of $n$ reacting species the number of independent reactions is at most equal to $(n-1)$ see also Ex. 2] but only a number $r \leq (n-1)$ of them. In this case only $r$ relations will be independent.

Example 2

To describe the $H_2 - O_2$ kinetics the following set of equations is given :

| | |
|---|---|
| (1) | $H + O_2 \rightleftharpoons OH + O$ |
| (2) | $O + H_2 \rightleftharpoons OH + O$ |
| (3) | $2OH \rightleftharpoons H_2O + O$ |
| (4) | $H_2O + M \rightleftharpoons OH + H + M$ |
| (5) | $O_2 + M \rightleftharpoons 2O + M$ |
| (6) | $H_2 + OH \rightleftharpoons H_2O + H$ |
| (7) | $H_2 + M \rightleftharpoons 2H + M$ |
| (8) | $OH + M \rightleftharpoons O + H + M$ |

where M is a component which does not react but enters as a third body in the reactions 4,5,7,8. [When this third body is taken as a molecular Nitrogen, $N_2$, this same set of equations can be used to describe the processes of $H_2$ combustion in air .]

By taking as reference quantity the molecular mass (m) of the atomic hydrogen, the matrix $\underline{\underline{N}}$ formed by the coefficients $\bar{\nu}_{\ell i}$ is

| | 1 [H] | 2 [O] | 3 [$H_2O$] | 4 [OH] | 5 [$O_2$] | 6 [$H_2$] |
|---|---|---|---|---|---|---|
| (1) | -1 | 16 | 0 | 17 | -32 | 0 |
| (2) | 1 | -16 | 0 | 17 | 0 | -2 |
| (3) | 0 | 16 | 18 | -34 | 0 | 0 |

(2.5a)

$$
\begin{array}{c} (2.5b) \end{array}
\begin{array}{c|cccccc}
④ & 1 & 0 & -18 & 17 & 0 & 0 \\
⑤ & 0 & 32 & 0 & 0 & -32 & 0 \\
⑥ & 1 & 0 & 18 & -17 & 0 & -2 \\
⑦ & 2 & 0 & 0 & 0 & 0 & -2 \\
⑧ & 1 & 16 & 0 & -17 & 0 & 0
\end{array} = \underline{\underline{N}}
$$

where each row corresponds to a reaction and each column to a component of the mixture, as shown on the side and on the top of the matrix. The column corresponding to M has been omitted: all its elements are equal to zero since, as said, M does not react.

The number of reacting components is 6, the number of atomic species is 2, hence only 6–2 = 4 reactions are independent. The others are combinations of the first four which can be symbolically denoted as:

$$
\begin{aligned}
⑤ &= ① + ③ + ④ \\
⑥ &= ② + ③ \\
⑦ &= ② + ③ + ④ \\
⑧ &= ③ + ④
\end{aligned} \qquad (2.6)
$$

meaning, for instance, that the reaction (6) is (formally) obtained by adding reactions (2) and (3).

We can then write the matrix $\underline{\underline{N}}$ in the partitioned form

$$\underline{\underline{N}} = \begin{bmatrix} \underline{\underline{N}}_1 \\ \underline{\underline{N}}_2 \end{bmatrix} \tag{2.7}$$

where the (4x6) matrix $\underline{\underline{N}}_1$, is the submatrix of $\underline{\underline{N}}$ formed by its first four rows and the (4x6) matrix $\underline{\underline{N}}_2$ is the submatrix of $\underline{\underline{N}}$ formed by its last four rows. Since the rows of $\underline{\underline{N}}_2$ are linear combinations of the rows of $\underline{\underline{N}}_1$, we can also write:

$$\underline{\underline{N}}_2 = \underline{\underline{B}} \cdot \underline{\underline{N}}_1 \tag{2.8}$$

where, upon equations (2.6), the fourth order squared matrix $\underline{\underline{B}}$ is given by:

$$\underline{\underline{B}} = \begin{bmatrix} 1 & 0 & 1 & 1 \\ 0 & 1 & 1 & 0 \\ 0 & 1 & 1 & 1 \\ 0 & 0 & 1 & 1 \end{bmatrix}$$

In terms of the elements of the matrices $\underline{\underline{N}}$, $\underline{\underline{N}}_1$ and $\underline{\underline{N}}_2$ the mass conservation equations (2.3) become:

$$\sum_{i=1}^{6} (\underline{\underline{N}})_{\ell i} = 0 \quad ; \qquad (1 \leq \forall \ell \leq 8)$$

or, on account of eq. (2.7) and (2.8) :

$$\sum_{i=1}^{6} (\underline{\underline{N}}_1)_{ri} = 0 \quad ; \qquad (1 \leq \forall r \leq 4)$$

$$\sum_{i=1}^{6} (\underline{\underline{N}}_2)_{ri} = \sum_{i=1}^{6} \sum_{k=1}^{4} (\underline{\underline{B}})_{rk} (\underline{\underline{N}}_1)_{ki} =$$

$$= \sum_{k=1}^{4} \left[ (\underline{\underline{B}})_{rk} \sum_{i=1}^{6} (\underline{\underline{N}}_1)_{ri} \right]$$

and it is seen that only the first four, are independent statements.

Denote by $\delta \underline{c}$ and $\delta \underline{\xi}$ the (6x1) and (8x1) column matrices whose elements are the $\delta c_i (1 \leq i \leq 6)$ and $\delta \xi_j (1 \leq j \leq 8)$ respectively, see eq. (2.11) and partition $\delta \xi$ as :

$$\delta \xi = \begin{bmatrix} \delta \xi_1 \\ \delta \xi_2 \end{bmatrix}$$

where $\xi_{1_j}$ and $\xi_{2_j} (1 \leq j \leq 4)$ are the progress variables of the first and last four reactions (2.4), respectively Then, if $\tilde{\underline{\underline{A}}}$ denotes the transpose of the matrix $\underline{\underline{A}}$, the relations (2.11) become, on account of eq. (2.8) :

$$\delta \underline{c} = \underline{\underline{\tilde{N}}} \cdot \delta \underline{\xi} = \underline{\underline{\tilde{N}}}_1 \cdot \left[\delta \underline{\xi}_1 + \underline{\underline{\tilde{B}}} \cdot \delta \underline{\xi}_2\right] = \underline{\underline{\tilde{N}}}_1 \cdot \delta \underline{\eta} \tag{2.9}$$

where :

$$\delta \underline{\eta} = \delta \underline{\xi}_1 + \underline{\underline{\tilde{B}}} \cdot \delta \underline{\xi}_2$$

or, explicitly :

$$\begin{aligned} \delta \eta_1 &= \delta \xi_1 + \delta \xi_5 \\ \delta \eta_2 &= \delta \xi_2 + \delta \xi_6 + \delta \xi_7 \\ \delta \eta_3 &= \delta \xi_3 + \delta \xi_5 + \delta \xi_6 + \delta \xi_7 + \delta \xi_8 \\ \delta \eta_4 &= \delta \xi_4 + \delta \xi_5 + \delta \xi_7 + \delta \xi_8 \end{aligned} \tag{2.10}$$

The parameters $\eta_i$ are the four independent "progress variables" which describe "globally" the progress. From the "equilibrium thermodynamic" point of view they are the only relevant parameters since the virtual displacements of the concentrations $\delta c_i (1 \leq i \leq 6)$ are uniquely determined when the $\delta \eta_i$ are arbitrarily described. From the chemical kinetics point of view, however, they are not sufficient to properly describe the kinetics of the processes. To begin with, the "chemical reaction", whose "progress variable" is $\eta_i$, is a "fictitious" reaction which does not correspond to any chemically appropriate scheme. For instance, from eqs. (2.4) and (2.10) one finds that $\eta_1$ is the progress variable of the "reaction" :

$$H + 2O_2 + M \rightleftharpoons 3O + OH + M$$

which "makes no sense" from the chemical-kinetics point of view. On the other hand, speaking in terms of rates, the kinetics is properly described only when one prescribes (for instance) the forward rate and the equilibrium constant for each one of the 8 reactions (2.4), [although only 4 of these equilibrium constants are independent. (Ref.1)].

Thus in general, the number of constraints on the $\delta c_i'$ equals the number (r) of independent reactions. Specifically, as readily checked, the virtual displacements $\delta c_i$ must be such that :

$$\text{(2.11)} \qquad \delta c_i = \sum_{\ell=1}^{r} \nu_{\ell i} \delta \xi_\ell \; ; \quad (1 \leq \forall i \leq n-1)$$

where the displacements $\delta \xi_\ell$ are now completely arbitrary. The parameters $\xi_\ell$ are referred to as "progress variables" for the reactions. For "closed" systems they suffice to characterize uniquely the composition of the evolving mixture. When processes of mass diffusion cannot be neglected, the elementary fluid particle constitutes an open system and the progress variables are no-longer sufficient to characterize its composition.

Example 3

In most of our examples we shall choose a particular composite system(which shall be referred to as "typical" system) which is sufficiently simple as to avoid lengthy or cumbersome formulae yet it is significant enough as both "typical" processes, exchange of energy, may occur among its sub-systems.

This "typical" system is a gas mixture consisting of three sub-systems : 1) the ensemble of the translational and rotational degrees of freedom of the molecules of a biatomic gas; 2) the ensemble of the vibrational degree of freedom of the molecules; 3) the corresponding atomic gas. Mass exchange may occur among sub-systems (1) and (3); energy exchange may occur among sub-systems (1) and (2). The subsystems (1) and (3) are in mutual equilibrium with respect to energy exchange processes.

Thus, in the "typical" system $n = 2, k = 1, d = 4$ and a state of the system is characterized by the values of $\mathcal{V}, S, M, M_1, S_v$ where : $M_1$ is the mass of the atomic gas, $S_v$ the entropy of the ensemble of the vibrational degree of freedom of the molecules; $M = M_1 + M_2$ and $M_2$ is the mass of the molecular gas.

For the "typical" system the chemical reaction is the dissociation of the biatomic molecule :

$$[c_2] \rightleftarrows 2[c_1]$$

hence by dropping the subscript $\ell$ (since there is only one reaction), and by noticing that

$$\bar{\nu}_1 = 0 \quad \bar{\nu}_2 = 1 \quad \bar{\nu}_1' = 2 \quad \bar{\nu}_2' = 0$$

(2.12) $$M_r = M_2 = 2\,M_1$$

$$\nu_1 = \frac{2\,M_1}{M_2} = 1 \qquad \nu_2 = -\frac{M_2}{M_2}$$

Clearly in this simple case $\xi$ can be taken to be the atoms mass concentration "c".

We can now discuss the possible equilibrium conditions of the subject mixture [the analyses shall be limited to the case that the mixture remains in a single phase]. A necessary condition is, upon eq. (2.11):

$$\sum_{l=1}^{r} A_l \, \delta\xi_l + \sum_{j=1}^{k} (\Theta_i - T)\delta s_j = 0 \tag{2.13}$$

where the quantities $A_l$ defined by :

$$A_l = \sum_{i=1}^{n-1} \nu_{li}(g_i - g_n) = \sum_{i=1}^{n} \nu_{li} g_i \tag{2.14}$$

are known as the "de-Donder" affinities for the reactions and can be considered as the intensive parameters conjugated to the progress in the energy representation. They also give the change in the specific Gibbs free energy (see later on) due to the reaction..

Example 4

For the "typical" system, one has, upon, equations (2.12):

$$A = \nu_1 g_1 + \nu_2 g_2 = g_1 - g_2 \tag{2.15}$$

i.e., as expectable, the affinity is just the difference between the electrochemical potentials of the atom and the molecule.

For each process ("r" mass exchanges and "k" energy exchanges) there are two possible "forms" of equilibrium. Unconstrained equilibrium prevails when $\delta\xi_\ell$ (resp. $\delta s_i$) is different from zero (absences of constraints) and the corresponding "chemical" (resp. "thermal") equilibrium condition is:

$$A_\ell = 0 \qquad [\Theta_i = T]$$

Each such relation reduces by one unity the number of degrees of freedom of the isolated system.

Constrained equilibrium prevails when $\delta\xi_\ell$ (resp. $\delta s_i$ ) is zero i.e. when mass (resp. energy) exchange between the subsystem is prevented.. In this case no limitation is imposed on the conjugated intensive parameter : if it is different from zero the constraint is active, if it is equal to zero the constraint is non-active.

Thus the composite system may find itself in a large variety of different types of equilibrium depending on the number and type of equilibrium of the different processes which may take place among the sub-systems.

By reverting to the general notation introduced in paragraph (1) we shall refer to as states of partial unconstrained equilibrium of order $(t)$ the states in which :

$$y_j(s, v, x_i) = 0 \qquad 1 \le j \le t \tag{2.16}$$

By suitably rearranging the correspondence between the ordered set $(1, 2 \ldots\ldots t, t+1, \ldots\ldots d-2)$ and the physical variables we can then define a state of partial equilibrium of order "$t$" as a state in which the first "$t$" intensive parameters (other than $T$ and $p$, naturally) vanish identically. Clearly, there are $\binom{d-2}{t}$ different types of partial equilibrium of order "$t$" each corresponding to a different set of "$t$" vanishing intensive parameters.

A state of full (complete) unconstrained equilibrium corresponds to $t = (d-2)$ i.e. to the vanishing of all intensive parameters other than $T$ and $p$.

Similarly, we shall define, the states of partial constrained equilibrium of order "$m$" as the states for which, in the isolated system, the last "$m$" specific variables (other than "$s$" and "$v$") are constant. There are $\binom{d-2}{m}$ different types of partial constrained equilibrium of order "$m$".

Any of the possible equilibrium states of the system is characterized by a value of $t$ $(0 \le t \le d-2)$ and of $m$ $(0 \le m \le d-2)$. In particular, when $t + m = d-2$ the system is in a state of constrained complete equilibrium.

Upon the shifting equilibrium hypothesis, the

number of specific degrees of freedom of the subject system during its evolution depends on the type of equilibrium which prevails during the evolution itself. Specifically :

i) if no equilibrium prevails the system has "d" specific degrees of freedom,

ii) if a "pure" partial equilibrium of order "t" prevails $(m = 0)$ the system has $(d-t)$ specific degrees of freedom, (in particular, only 2 in conditions of full unconstrained equilibrium),

iii) if a pure partial constrained equilibrium of order "m" prevails $(t=0)$ the system has "d" [resp. $(d-m)$] specific degrees of freedom according to whether during the evolution it can be considered open [resp. closed],

iv) for "t" and "m" both different from zero, the number of specific degrees of freedom is obtained by combining the statement in ii) and iii) above.

Notice that :

a) A condition of partial constrained equilibrium can persist indefinitely only when the acting constraints are perfect. In practice, since one is only interested in the conditions of a system during a prescribed period of observation a constraint acting between two subsystems may be considered perfect if during the period of observation the extensive quantity exchanged between them is a negligible percentage of the total quantity present in them (*). Thus air in normal sea level conditions can be considered to be in a state of constrained equilibrium both with respect to mass exchange (no chemical reactions) and energy exchange (no excitation of the internal degrees of freedom of the constituent molecules).

b) macroscopic equilibrium thermodynamic can only tell the consequences stemming from the fact that a certain type of equilibrium prevails, it cannot say when, and under which conditions such equilibrium occurs. Information concerning this latter point must come from "outside" equilibrium thermodynamics and must involve considerations connected with the characteristic times of the processes as compared with the observation times and (when the system is no longer isolated) with the times associated with a) changes in the environment of the system and b) the interaction between the system and its environment.

---

(+) Thus whether or not a constraint can be considered perfect depends not only on its nature but also on two factors : i) the period of observation, ii) the prescribed accuracy. These considerations always apply when practical situations must be related to "theoretical models".

Example 5

For the "typical" system there are :

a) $\binom{d-2}{1} = \binom{4-2}{1} = 2$ different first order partial

equilibria defined by :

1) Chemical equilibrium $\Longrightarrow$ $A = 0$ $\left[g_1 = g_2\right]$

2) Thermal equilibrium $\Longrightarrow$ $\Theta = T$

b) two different first order partial constrained equilibrium corresponding to the cases in which : 1) the dissociation is frozen, 2) the energy exchange is frozen

c) one complete unconstrained equilibrium defined by :

$$A = 0 \qquad \Theta = T$$

d) one completely constrained equilibrium corresponding to the case in which both the dissociation and the exchange of energy are frozen.

Considering the case in which mass diffusion cannot be neglected, the subject system may evolve in any of the following conditions :

i) complete non-equilibrium : four specific degrees of

freedom (s.d.f) : $s, v, c, s_v$ may change because of the interaction with the environment and because of the "internal" processes;

ii) chemical equilibrium : three s.d.f. which reduce to two in the case that the vibrational degree of freedom is frozen [first condition of complete constrained equilibrium] (upon the plausible assumption that subsystem (2) cannot interact directly with the environment)

iii) thermal equilibrium : three s.d.f. which, in this case, do not reduce to two even when the dissociation process is frozen [second condition of complete constrained equilibrium] since the atoms concentration may change upon <u>direct</u> interaction with the environment (i.e. mass diffusion)

iv) chemical and thermal equilibrium : two s.d.f.

v) completely constrained equilibrium : three s.d.f. since one subsystem does not interact directly with the environment whereas the others do.

Each of these conditions correspond to well identifiable situations which are often met in practice.

Upon the hypothesis of shifting equilibrium the differential of the extensive variables are certainly virtual displacements, it then follows that the energy function has the following additional property.

"The function $u = u(s, v, x_i)$ has the relative minimum in a condition of partial equilibrium of order "$t$" ; this minimum decreases as the order ofthe equilibrium increases and attains its lowest value in conditions of full unconstrained equilibrium".

Example 6

As it will be shown in the next paragraph, if the system is stable the equations defining the conditions of unconstrained equilibrium can be solved to obtain the values of partial equilibrium of the corresponding specific variables. Hence if one lets :

$$(c)_{pe} = (c)_{A=0} = (c)_{pe}(s, v, s_v)$$

$$(s_v)_{pe} = (s_v)_{T=\Theta} = (s_v)_{pe}(s, v, c)$$

$$c_e = (c)_{A=0,\,\Theta=T} = c_e(s, v)$$

$$(s_v)_e = (s_v)_{A=0,\,\Theta=T} = (s_v)_e(s, v)$$

one has the following inequalities :

$$u(s, v, s_v, c) > u\left[s, v, (c)_{pe}, s_v\right] > u\left[s, v, c_e, (s_v)_e\right]$$

$$\begin{bmatrix} \forall s, v, s_v \\ c = (c)_{pe} \end{bmatrix} \qquad \begin{bmatrix} \forall s, v \\ c = c_e \,;\; s_v = (s_v)_e \end{bmatrix}$$

$$u(s, v, c, s_v) > u\left[s, v, c, (s_v)_{pe}\right] > u\left[s, v, c_e, (s_v)_e\right]$$

$$\begin{bmatrix} \forall s, v, c \\ s_v = (s_v)_{pe} \end{bmatrix} \qquad \begin{bmatrix} \forall s, v \\ c = c_e \,;\; s_v = (s_v)_e \end{bmatrix}$$

where below each inequality the conditions under which it holds have been reported.

## 3. THERMODYNAMIC STABILITY

We have discussed so far only the necessary condition required by the equilibrium postulate.

The sufficient conditions $\delta^2 U > 0$ implies a further property of the internal energy function for thermodynamically stable systems . We shall consider only single phase systems which are, therefore, thermodynamically stable.

The stability conditions imply that the matrix formed by the second derivatives of the function $u = u(s, v, x_i)$ is positive definite.

The necessary and sufficient conditions for the

positive definite character of a matrix lead to a number of significant inequalities among certain types of thermodynamic derivatives. These inequalities can be summarized in two "stability rules" involving the couples conjugated with respect to the specific internal energy (*).

In a system with "d" specific degrees of freedom any thermodynamic derivative involving intensive and/or specific quantities is uniquely defined only when one assigns the $(d-1)$ variable held constant in the differentiation process. With specific reference to the energy representation given by equation (1.6) these variables can be chosen from among the set of $(2d+1)$ quantities constituted by the specific internal energy, its "d" specific natural variables and their "d" conjugate parameters. A set $z_i$ of (d-1) such quantities will be referred to as an allowable set if it does not contain: a) any couple (or any combination) of parameters conjugated with respect to the internal energy, b) the internal energy itself(**).

(*) Also the matrix formed by the second derivatives of the density of energy $u^+ = \varrho u = u^+(\varrho s, \varrho v, \varrho x_i)$ (where $\varrho$ is the mass density) is positive definite. This condition leads to inequalities among other classes of thermodynamic derivatives. Still others can be deduced by introducing the couples conjugated with respect to the entropy. None of them will be considered here.

(**) Other thermodynamic quantities will be defined later on, namely the thermodynamic potentials. They are also to be excluded from an allowable set.

Example 7

For the typical system the sets $z_i$ contain three vari -ables. The following sets are allowable sets:

$$(s, v, c)\ ;\ (s, v, s_v);$$

$$(T, v, c)\ ;\ (v, c, s_v)$$

$$[T, p, (g_1 - g_2)]\ ;\ [v, (g_1 - g_2), (\Theta - T)]$$

The following sets are not allowable sets:

$$(T, s, x)\ ;\ (x, p, v);\ [x, (g_1 - g_2), c]$$

$$[x, (\Theta - T), s_v]\ ;\ [u, x, y];\ [f(T, s), x, y]$$

where $f(T, s)$ is any arbitrary combination of $T$ and $s$, and $x, y$ are any two thermodynamic quantities.

First stability rule. "The derivative of any intensive parameter with respect to its specific conjugate parameter is always positive provided the quantities held constant in the differentiation process consitute an allowable set"

The first rule implies that:

i) For any couple of conjugate parameters one can define $2^{(d-1)}$ derivatives which are necessarily positive upon the thermodynamic stability of the system (these derivatives are, however, not necessarily distinct or independent).

Example 8

In the "typical" system $d=4$ and one can define $2^3=8$ positive derivatives for each conjugate couple. The allowable sets for the derivative $\partial T/\partial s$ are the four sets:

$$(c, v, s_v) ; \left[v, (g_1 - g_2), s_v\right] ; \left[v, c, (\Theta - T)\right] \tag{3.1}$$

$$\left[v, (g_1 - g_2), (\Theta - T)\right]$$

plus the other four obtained by replacing $v$ with $p$.

ii) The derivatives involving: a) non-conjugate parameters, b) two extensive parameters; c) two intensive parameters and the derivatives of two conjugate parameters for which the variables held constant do not constitute an allowable set, are not "effected" by the stability of the system. Hence they can either have any sign (depending on the particular system considered and/or on its state) or have a sign which follows from their definition and the Gibbs' relation (in one of its several forms, see later on).

Example 9

For the "typical" system derivatives such as

$$\left(\frac{\partial T}{\partial v}\right)_{z_i} \; ; \quad \left[\frac{\partial (g_1 - g_2)}{\partial s}\right]_{z_i}$$

where $z_i$ is any set of three state parameters, can have any sign. Furthermore

$$\left(\frac{\partial s}{\partial v}\right)_{u,c,s_v} = \frac{P}{T} > 0 \; ; \quad \left(\frac{\partial s}{\partial c}\right)_{u,v,s_v} = -\frac{A}{T} \gtreqless 0$$

where subscripts denote quantities held constant in the differentiation process and the equalities follows from the Gibbs relation (1.8).

Two derivatives A and B of the type mentioned in the first stability rule are said to be homologous when the set of variables held constant in the definition of A can be obtained from that of B by replacing one variable with its conjugate. The second stability rule leads to inequalities among homologous derivatives. Specifically:

Second stability rule. "Given two homologous derivatives, the one in which the intensive parameter is held constant can never be larger than the other". Conversely, the ratio of two non-homologous derivatives is not effected by the stability conditions and can be greater, equal or smaller than one, depending on the system and its state.

Example 10

For the "typical" system, thermodynamic stability implies that:

$$\left(\frac{\partial T}{\partial s}\right)_{v,c,s_v} \geq \left(\frac{\partial T}{\partial s}\right)_{p,c,s_v} \geq \left(\frac{\partial T}{\partial s}\right)_{p,(g_1-g_2),s_v} > 0$$

(3.2)
$$\left(\frac{\partial T}{\partial s}\right)_{v,c,s_v} \geq \left(\frac{\partial T}{\partial s}\right)_{v,c,(\Theta-T)} > 0$$

$$\left[\frac{\partial(g_1-g_2)}{\partial c}\right]_{s,v,s_v} \geq \left[\frac{\partial(g_1-g_2)}{\partial c}\right]_{T,v,s_v} > 0$$

On the other hand:

$$0 < \left(\frac{\partial T}{\partial s}\right)_{p,c,s_v} \gtreqless \left(\frac{\partial T}{\partial s}\right)_{v,c,(\Theta-T)} > 0$$

since these two derivatives are not homologous.

As shown by the first two inequalities in the example the second rule can be applied repeatedly and one arrives at the following statement:

"Any derivatives of an intensive parameter with respect to its specific conjugate parameter attains its smallest (resp. largest) value when the parameters held constant are all intensive (resp. all extensive). (If these two values coincide all different derivatives are equal)".

Notice that the above statement does not imply that there are not still smallest (resp. largest) derivatives that can certainly be defined for any system, provided it is stable.

Denote by $(y_i, x_i)$; $(y_j, x_j)$ two couples of conjugate parameters (with $y_i, y_j$ intensive parameters) and by $z$ any allowable set of $(d-2)$ parameters different from $(y_i, y_j)$; $(x_i, x_j)$. From the relation:

$$y_i = y_i\left[x_i, y_j(x_i, x_j, z), z\right]$$

it follows that

$$\left(\frac{\partial y_i}{\partial x_i}\right)_{x_j,z} = \left(\frac{\partial y_i}{\partial x_i}\right)_{y_j,z} + \left(\frac{\partial y_i}{\partial x_j}\right)_{x_i,z}\left(\frac{\partial x_j}{\partial y_j}\right)_{x_i,z}\left(\frac{\partial y_j}{\partial x_i}\right)_{x_j,z}$$

On the other hand it can be proved that: (*) (see foll. page)

$$\left(\frac{\partial y_i}{\partial x_j}\right)_{x_i,z} = \left(\frac{\partial y_j}{\partial x_i}\right)_{x_j,z}$$

It follows then that the two homologous derivatives $(\partial y_i / \partial x_i)_{x_i,z}$ and $(\partial y_i / \partial x_i)_{y_j,z}$ are equal if and only if

$$(3.3) \qquad 0 = \left(\frac{\partial y_i}{\partial x_j}\right)_{x_i,z} = \left[\left(\frac{\partial y_j}{\partial x_i}\right)_{x_j,z}\right]$$

Example 11

The first, second, third and fourth equalities in (3.2) are valid if and only if, respectively (recall that the conjugate of $v$ is $(-p)$

$$\left(\frac{\partial T}{\partial v}\right)_{s,c,s_v} = -\left(\frac{\partial p}{\partial s}\right)_{v,c,s_v} = 0$$

$$\left(\frac{\partial T}{\partial c}\right)_{s,p,s_v} = \left[\frac{\partial (g_1 - g_2)}{\partial s}\right]_{c,p,s_v} = 0$$

(*) This set of equalities belong to the so-called Maxwell's equalities. It can be readily proved if one introduces the notion of generalized potential. Since the set $(x_i, x_j, z)$ is, by definition, an allowable set, it is possible to find a thermodynamic potential $\varphi = \varphi(x_i, x_j, z)$ such that

$$y_i = \partial\varphi / \partial x_i \quad ; \quad y_j = \partial\varphi / \partial x_j$$

Then the equality (3.3) follows from the continuity of $\varphi$ and of its derivatives.

$$\left(\frac{\partial T}{\partial s_v}\right)_{s,v,c} = \left[\frac{\partial(\Theta - T)}{\partial s}\right]_{v,c,s_v} = 0$$

$$\left[\frac{\partial(g_1 - g_2)}{\partial s}\right]_{c,v,s_v} = \left[\frac{\partial T}{\partial c}\right]_{s,v,s_v} = 0$$

## 4. CONSEQUENCES OF THE STABILITY

### 1. Equilibrium value of extensive variables

Let again $(y_i, x_i)$ denote a couple of parameters conjugate with respect to the internal energy. As said, a condition of partial equilibrium of order $(t)$ is defined by the set of $(t)$ equations:

$$y_i = y_i(s, v, x_j) = 0 \qquad (1 \leq i \leq t)$$

This set of equations defines implicitly the "partial equilibrium" values of the first $t$ variables $(x_i)_{pe}$.

A necessary and sufficient condition for the obtainment of unique values of the $(x_i)_{pe}$ from these relations is that the Jacobian $\partial(y_i)/\partial(x_i)$ be different from zero. But this

Jocobian is a principal minor of the stability matrix and, therefore, it is essentially positive for stable systems. Hence, as a consequence of the stability, in conditions of partial equilibrium of order $t$ one has:

$$(x_i)_{pe} = x_i\left[s, v, x_j\right] ; \qquad \left[1 \leq i \leq t ; j \neq i\right]$$

2. Stability measures

Any of the derivatives mentioned in the first stability rule can be taken as a measure of the thermodynamic stability of the system with respect to well defined changes of its state.

Some such classes of stability measure are given specific names.

A) Stability with respect to entropy changes. Specific heat coefficients.

The coefficients of specific heat are defined as:

$$\frac{1}{c_{z_i}} = \frac{1}{T}\left(\frac{\partial T}{\partial s}\right)_{z_i} > 0 \tag{4.1}$$

where $z_i$ is any allowable set of $(d-1)$ thermodynamic quantities and the inequality follows from the first stability rule.

Two classes of specific heat coefficients can be defined. The constant volume $c_{v,w_i}$ and the constant pressure $c_{p,w_i}$ specific heat coefficients where, now, $w_i$ denotes any allowable set of $(d-2)$ quantities:

$$\frac{T}{c_{v,w_i}} = \left(\frac{\partial T}{\partial s}\right)_{v,w_i} ; \quad \frac{T}{c_{p,w_i}} = \left(\frac{\partial h}{\partial T}\right)_{p,w_i} \tag{4.2}$$

Equivalently, accounting for the Gibbs relations in the energy and enthalpy (see following paragraph) representations of the system the specific heats can also be defined as:

$$c_{v,w_i} = \left(\frac{\partial u}{\partial T}\right)_{v,w_i} ; \quad c_{p,w_i} = \left(\frac{\partial h}{\partial T}\right)_{p,w_i} \tag{4.3}$$

The following remarks are important:

- The specific heat coefficients are a measure of the thermodynamic stability of the system with respect to entropy changes at $w_i = \text{const.}$
- In a system with (d) specific degrees of freedom one can define $2^{(d-2)}$ "types" of constant volume and constant pressure specific heat coefficients corresponding to the $2^{(d-2)}$ possible different allowable sets $w_i$. These specific heats are not necessarily distinct or independent.
- For any given $w_i$ the second stability rule implies that:

$$c_{p,w_i} \geq c_{v,w_i} \qquad (\forall w_i) \tag{4.4}$$

- the equality holding if and only if [see eq. (3.3)]:

$$(4.5) \qquad \left(\frac{\partial T}{\partial v}\right)_{s,\mathfrak{w}_i} \equiv -\left(\frac{\partial p}{\partial s}\right)_{v,\mathfrak{w}_i} = 0$$

- If two compatible sets $\mathfrak{w}_1$ and $\mathfrak{w}_2$ are such that $\mathfrak{w}_1 \equiv (x_j, \omega_i)$ $\mathfrak{w}_2 \equiv (y_j, \omega_i)$ with $\omega_i$ any compatible set of $(d-3)$ variables, it is always (for stable systems)

$$(4.6) \qquad \begin{aligned} c_{v,y_i,\omega_i} &\geq c_{v,x_i,\omega_i} \\ c_{p,y_j,\omega_i} &\geq c_{p,x_j,\omega_i} \end{aligned}$$

whereas

$$(4.7) \qquad \frac{c_{v,y_i,\omega_i}}{c_{p,x_j,\omega_i}} \gtrless 1$$

depending on the system and its state.

The two constant volume specific heat coefficients are equal if and only if:

$$\left(\frac{\partial T}{\partial x_j}\right)_{s,p,\omega_i} \equiv \left(\frac{\partial y_j}{\partial s}\right)_{p,x_j,\omega_i} = 0$$

B. <u>Stability with respect to volume changes. Characteristic speeds</u>.

The thermodynamic characteristic speeds are defined as:

$$a^2_{z_i} = -v^2 \left(\frac{\partial p}{\partial v}\right)_{z_i} = \left(\frac{\partial p}{\partial \varrho}\right)_{z_i} > 0$$

It proves useful to define two classes of characteristic speeds: the Laplacian (or isentropic) speeds $a_{s,w_i}$ (frequently called "speeds of sound") and the Newtonian (or isothermic) characteristic speeds $a_{T,w_i}$:

$$a^2_{s,w_i} = \left(\frac{\partial p}{\partial \varrho}\right)_{s,w_i}$$
$$a^2_{T,w_i} = \left(\frac{\partial p}{\partial \varrho}\right)_{T,w_i} \tag{4.8}$$

They are a measure of the stability of the system with respect to isentropic (resp. isothermal) volume changes at $w_i$=const.

One can repeat here the same remarks made in connection with specific heat. Thus one has:

$$a_{s,w_i} \geq a_{T,w_i} \qquad (\forall\, w_i)$$

$$a_{s,x_j,\omega_i} \geq a_{s,y_j,\omega_i} \quad (\forall\, \omega_i)$$

$$a_{T,x_j,\omega_i} \geq a_{T,y_i,\omega_i} \tag{4.9}$$

$$\frac{a_{s,x_j,\omega_i}}{a_{T,y_j,\omega_i}} \gtreqless 1$$

with the first three equalities holding if, and only if,

respectively

$$\left(\frac{\partial p}{\partial s}\right)_{v, w_i} = 0$$

(4.10) $$\left(\frac{\partial p}{\partial x_j}\right)_{s, y_j, \omega_i} = 0$$

$$\left(\frac{\partial p}{\partial x_j}\right)_{T, y_j, \omega_i} = 0$$

It is seen that the condition for the validity of the equality (4.9) coincides with that for the validity in (4.4). This is so, because, quite generally, if

(4.11) $$\gamma_{w_i} = \frac{c_{p,w_i}}{c_{v,w_i}} \geq 1$$

is the specific heat ratio corresponding to an allowable set $w_i$, one has, upon the definitions (4.8), (4.3) and the relation (4.11)

(4.12) $$\gamma_{w_i} = \frac{a^2_{s, w_i}}{a^2_{T, w_i}} .$$

C.- Stability with respect to changes in composition

This type of stability is measured by the classes of derivatives:

$$\left[\frac{\partial(g_r-g_n)}{\partial c_r}\right]_{z_i} > 0 \qquad (1 \le \forall r \le n-1) \tag{4.13}$$

for which no terminology is available. Each such derivative is a measure of the stability of the system with respect to a change in the concentration $c_r$ at $z_i$ = const. Chains of inequalities similar to those discussed for the specific heats and the characteristic speeds can be derived for these quantities and will not be formulated explicitly.

By recalling that the affinity $A_j$ is the parameter conjugated to the progress variable one deduces that the quantities:

$$\left(\frac{\partial A_r}{\partial \xi_r}\right)_{z_i} > 0 \tag{4.14}$$

are the measure of the stability of the system with respect to the r-th reaction (i.e. with respect to the changes in composition induced by the r-th reaction) occuring at $z_i$ = const.

D. - <u>Stability with respect to the internal energy exchanges</u>

This type of stability is measured by the derivatives:

$$(4.15) \qquad \left[\frac{\partial(\Theta_r - T)}{\partial s_r}\right]_{z_i} > 0 \qquad (\forall r, 1 \le r \le k)$$

and to them one may apply all the considerations developed in connection with the other measures of stability.

Example 12

The definition of the four constant volume specific heat coefficients correspond to the four allowable sets given by (4.1). Likewise for the four constant pressure specific heat.

The definitions of the four Laplacian characteristic speeds correspond to the following four allowable sets:

$$(s, c, s_v)\ ;\ (s, A, s_v);\ \left[s, c, (\Theta - T)\right];\ \left[s, A, (\Theta - T)\right]$$

In a system with only three specific degrees of freedom [say: a dissociating biatomic gas] $d=3$ and there are only two Laplacian characteristic speeds:

$$a^2_{s,1} = \left(\frac{\partial p}{\partial \varrho}\right)_{s,c} \ ; \quad a^2_{s,2} = \left(\frac{\partial p}{\partial \varrho}\right)_{s,A} \le a^2_{s,1}$$

They are often referred to as 'frozen" and "equilibrium"

speeds of sound. Notice explicitly that they are thermodynamically well defined quantities and the terminology "equilibrium speed of sound" has nothing to do with the "actual" value of the affinity which can well be different from zero. Similar remarks apply for the "equilibrium" specific heats and for systems with larger number of "degrees of freedom".

## 5. THERMODYNAMIC POTENTIALS

As we have seen, the natural variables of the internal energy are all extensive and the energy fundamental relation characterizes completely and uniquely the thermodynamic behaviour of the system. As a consequence, a particular state of the system can be identified by prescribing any set of independent variables not necessarily all extensive.

The rather obvious question then arises: is it possible to define thermodynamic quantities such that, when their functional dependence upon an arbitrary set of independent variables is given, the thermodynamic behaviour of the system is completely and uniquely characterized ?.

The answer to this question is positive provided the new set does not contain any couple of parameters conjugated with respect to the internal energy (or the entropy) representa -tion. We have already met such sets and, as it is recalled we have called them allowable sets. Hence any allowable set is a set of natural variables for a new thermodynamic quantity called a thermodynamic potential and the behaviour of the system is completely and uniquely characterized when the functional dependence of a thermodynamic potential upon its natural variables is known.

Quantities such as the enthalpy, the Helmotz free energy, the Gibbs free energy are well known examples of thermodynamic potentials. Since we shall also need to use less known potentials it proves useful to give a short outline of their general theory. Time limitations prevent a full treatment of the topic. Thus we shall consider only the thermodynamic potentials which are derived from the entropy (so called Mathieu functions). Furthermore, we shall not consider explicitly the class of potential whose natural variables include the conjugate of the total mass.

The (energy) thermodynamic potentials are defined as Legendre transforms of the internal energy. Specifically, we define a generalized potential of order $t$ as the partial Legendre transforms of order $t$ of the energy function with respect to $t$ of its natural variables. From this definition and from the properties of the Legendre transforms one can obtain all the relevant properties of the thermodynamic potentials.

For the convenience of the students, we shall offer two mnemonic rules that make it possible to readily derive all essential results concerning these potentials. By definition, any allowable set of variables can be obtained from the set of natural variables for the internal energy by replacing a number $q \geq 0$ of extensive parameters with the corresponding conjugate intensive parameters. We shall express this fact by saying that in any allowable set containing a number $q \geq 0$ of intensive variables the role of $q$ conjugated couples has been "interchanged" We can then formulate the first mnemonic rule as follows:

First mnemonic rule - Given an allowable set containing $q \geq 0$ intensive parameters, start from the Euler relation for the internal energy and bring to the left hand side the $q$ couples which have been interchanged. In the resulting equality the left hand side yields the definition of the corresponding thermodynamic potential and the right hand side gives its Euler relation.

Thus, for instance, if one wants the potentials corresponding to the following allowable sets:

$$(s, p, x_i) \ ; \ (T, v, x_i) \ ; \ (s, v, y_r, x_{i \neq r}) \ ; \ (T, p, x_i)$$

one starts from the specific Euler relation:

$$u = Ts - pv + \sum_{i=1}^{d-2} y_i x_i + g_n$$

and brings to the left hand side, respectively, the couples:

$$(p, v);\ (T, s);\ (y_r, x_r) \text{ and } \left[(T, s), (p, v)\right]$$

to obtain:

$$h = u + pv = Ts + \sum_i y_i x_i + g_n$$

$$f = u - Ts = -pv + \sum_i y_i x_i + g_n$$

$$\Psi_r = u - y_r x_r = Ts - pv + \sum_{i \neq r} y_i x_i + g_n$$

$$g = u - Ts + pv = \sum_i y_i x_i + g_n \,.$$

The potentials $h$, $f$, $g$, are, respectively, the well known specific enthalpy, and the specific Helmotz and Gibbs potentials. The potential $\Psi_r$ is a new one and no name has been given to it. The right hand sides of these equations give the Euler relations for the corresponding potentials. From these expressions one may readily understand the following property of any thermodynamic potential :

"A total generalized thermodynamic potential is a first order homogeneous function of its extensive natural variables".

The second mnemonic rule concerns the differentials of the total (or specific) thermodynamic potential:

Second mnemonic rule – To obtain the differential of a total (or specific) thermodynamic potential:

a) Start from the differential of the total (or specific) internal energy;

b) For each couple of conjugate parameters which has been interchanged, interchange the differential and change the sign.

Quite generally, if the number of extensive degrees of freedom of the system is $(d+1)$ and if we denote now by $x_i$ the value of any extensive parameter $x_i$ per unit mass, $y_i$ its intensive parameters conjugated in the energy representation and $g$ the parameter conjugated to the total mass, we can write:

Euler Relation $$u = \sum_{j=1}^{t} y_j x_j + \sum_{r=1}^{d-t} y_{t+r} x_{t+r} + g_n$$

$$(5.1)$$

Gibbs Relation $$du = \sum_{j=t}^{t} y_j dx_j + \sum_{r=1}^{d-t} y_{t+r} dx_{t+r}$$

For the potential $\Psi_t$ of order $t$:

$$\Psi_t = \Psi_t \left(y_j , x_{t+r}\right) ; \quad \left[1 \le j \le t ; 1 \le r \le (d-t)\right] \tag{5.2}$$

one has:

$$\Psi_t = u - \sum_{j=1}^{t} y_j x_j$$

Euler relation

$$\Psi_t = \sum_{r=1}^{d-t} y_{t+r} x_{t+r} + g_n \tag{5.3}$$

Gibbs relation

$$d\Psi_t = -\sum_{j=t}^{t} x_j \, dy_j + \sum_{r=1}^{d-t} y_{t+r} \, dx_{t+r} \, .$$

Notice that:

- Since the actual correspondence between the ordered set (1,2.., t ) and the (d) physical specific variables can be done in $\binom{d}{t}$ different ways, there are $\binom{d}{t}$ different specific potentials of order $t$.
- The highest order potential per unit mass is the (only) potential of order $d$. The corresponding Gibbs relation is nothing but the Gibbs-Duhem relation (referred to the unit mass of the system).
- From the Gibbs relation one obtains the definition of the thermodynamic quantities other than the natural variables as:

$$x_j = -\frac{\partial \Psi_t}{\partial y_j} = x_j(y_j \, , x_{t+r}) \qquad \begin{matrix}(1 \le j \le t)\\(1 \le i \le t)\end{matrix}$$

(5.4)

$$y_{t+r} = \frac{\partial \Psi_t}{\partial x_{t+r}} = y_{t+r}(y_j, x_{t+r}) \qquad (1 \le r \le d-t)$$

- The postulate of minimum discussed in section (2) for the internal energy applies to each potential $\Psi_t = \Psi_t(y_j \, , x_{t+r})$ $[1 \le j \le t; 1 \le j \le d-t]$ provided one replaces the words "isolated composite system" with the words "a composite system" in contact with a system of reservoirs holding its natural intensive variables constant" (open systems).

5.1.- Generalized enthalpy and generalized Helmotz potential

Among all such potentials we shall be especially interested in the potentials $\underset{\sim}{h}$ and $\underset{\sim}{f}$ which can be referred to as generalized enthalpy and generalized Helmotz potential, respectively. These generalizations are needed to derive, for mixtures which are not in thermal equilibrium. The generalized enthalpy $\underset{\sim}{h}$ is defined as the partial Legendre transform of the internal energy with respect to $v$ and to all entropies $s_j$: i.e.

$$\underset{\sim}{h} = u + pv - \sum_{j=1}^{k}(\Theta_j - T)s_j = Ts + \sum_{i=1}^{n-1}(g_i - g_n)c_i + g_n \tag{5.5}$$

The generalized Helmotz potential is defined as the partial Legendre transform of the internal energy with respect to all entropies i.e. with respect to $s$ and to all $s_j$ :

$$\underset{\sim}{f} = u - Ts - \sum_{j=1}^{k}(\Theta_j - T)s_j = -pv + \sum_{i=1}^{n-1}(g_i - g_n)c_i + g_n =$$

$$= \underset{\sim}{h} - pv - Ts \tag{5.6}$$

Hence:

$$\begin{aligned} \underset{\sim}{h} &= \underset{\sim}{h}\left[s, p, c_i, (\Theta_j - T)\right] \\ \underset{\sim}{f} &= \underset{\sim}{f}\left[T, v, c_i, (\Theta_j - T)\right] \end{aligned} \tag{5.7}$$

Notice that:

- In condition of thermal equilibrium the generalized enthalpy and Helmotz potential are equal to the "classical" enthalpy and Helmotz potential
- Since $\underset{\sim}{f}$ is defined as the partial Legendre transform with respect to <u>all</u> entropies an entirely equivalent set of natural variables for $\underset{\sim}{f}$ is the set $(T, v, c_i, \Theta_j)$ i.e.

$$f = f\left(T, v, c_i, \Theta_j\right) \tag{5.8}$$

- The two possible interpretations of $\underset{\sim}{f}$ lead to two entirely equivalent expressions for its Gibbs relation:

$$d\underset{\sim}{f} = -s\,dT - p\,dv + \sum_{i=1}^{n-1} (g_i - g_n)\,dc_i - \sum_{j=1}^{k} s_j\, d(\Theta_j - T)$$

$$d\underset{\sim}{f} = -s^{(e)} dT - p\,dv + \sum_{i=1}^{n-1} (g_i - g_n)\,dc_i - \sum_{j=1}^{k} s_j\, d\Theta_j$$

where:

$$s^{(e)} = s - \sum_{j=1}^{k} s_j$$

is the specific entropy of all subsystems in mutual thermal equilibrium. Obviously:

$$-s = \left(\frac{\partial \underset{\sim}{f}}{\partial T}\right)_{v, c_i, (\Theta_j - T)} \; ; \quad -s^{(e)} = \left[\frac{\partial \underset{\sim}{f}}{\partial T}\right]_{v, c_i, \Theta_j}$$

Example 13

For the 'typical' system the definition of generalized potentials $h$ and $f$ their corresponding Euler relations are: : : :

$$\underset{\sim}{h} = u + pv - (\Theta - T)s_v = h - (\Theta - T)s_v = Ts - g_2$$

$$\underset{\sim}{f} = u - Ts - (\Theta - T)s_v = u - Ts^{(e)} - \Theta s_v =$$

$$= f - (\Theta - T)s_v = -pv + g_2$$

Their Gibbs relations read:

$$d\underset{\sim}{h} = Tds + vdp - s_v d(\Theta - T)$$

$$d\underset{\sim}{f} = -sdT - pdv - s_v d(\Theta - T) =$$

$$= -s^{(e)}dT - pdv - s_v d\Theta$$

5.2. Partial specific quantities

Consider a composite system in thermal equilibrium. Any thermodynamic extensive quantity $Q$ can then be considered as a function of the allowable set $[T, p, m_i, (1 \le i \le n)]$. $Q$, by definition, is then first order homogeneous in the $M_i$. Upon the Euler theorem it then follows that:

$$Q = \sum_{i=1}^{n} \bar{Q}_i m_i$$

where

$$\bar{Q}_i = \left[\frac{\partial Q}{\partial m_i}\right]_{T,p,m_{j\neq i}}$$

The quantities $\bar{Q}_i$ are referred to as 'partial specific quantities". The specific quantity $q = Q/m$ where $m$ is the total mass of the system, can thus be expressed in terms of the partial specific quantities $\bar{Q}_i$ as:

$$q = \sum_{i=1}^{n} c_i \bar{Q}_i$$

The notion of partial specific quantity can be extended to the case of systems not in thermal equilibrium. In this case $Q$ must be considered function of the allowable set $[T; p; (\Theta_j - T), (1 \leq j \leq k); m_i, (1 \leq i \leq n)]$ so that the partial specific quantities must be defined as:

(5.9) $$\hat{Q}_i = \left[\frac{\partial Q}{\partial m_i}\right]_{T,p,\Theta_j,m_{r\neq i}}$$

and, clearly, it is again:

(5.10) $$Q = \sum_{i=1}^{n} \hat{Q}_i m_i \qquad q = \sum_{i=1}^{n} c_i \hat{Q}_i$$

In particular, for the specific entropy and the generalized enthalpy one has:

$$s = \sum_{i=1}^{n} c_i \hat{S}_i$$

$$h = \sum_{i=1}^{n} c_i \hat{\underset{\sim}{H}}_i$$

so that, substitution into equation (5.5) leads to the notable equality:

$$\hat{\underset{\sim}{H}}_i = T\hat{S}_i + g_i \qquad (5.11)$$

which generalizes the well known relation $\hat{H}_i = T\bar{S}_i + g_i$ valid in conditions of thermal equilibrium. From this equation, if one denotes by $(dg_i)_T$ the isothermic differential of $g_i$ one obtains:

$$T\,d\left(\frac{g_i}{T}\right) = (dg_i)_T - \frac{1}{T}\hat{\underset{\sim}{H}}_i\,dT. \qquad (5.12)$$

This equality also generalizes a well known thermodynamic equality valid in conditions of thermal equilibrium and will prove useful, later one, when discussing the expression for the entropy production.

## 6. THERMODYNAMIC MODELS

A thermodynamic model is prescribed when the explicit form of a fundamental relation is given.

As mentioned in the introduction these explicit forms cannot be obtained from purely macroscopic theories. One must resort to statistical thermodynamics and integrate its results with a number of suitable assumptions suggested by experimental observations.

A "thermodynamic model' may be used to describe, in particular ranges of the state parameters and to within some approximation which may or may not be known a priori, the thermodynamic behaviour of natural systems. It should be understood, however, that the choice of a model for a given natural system is related to the particular evolution of the system which one is interested in. This remark should appear almost self-evident after what has been said on the different types of equilibrium which may prevail during a given evolution. The same natural system may have to be described by different models for different evolutions.

### 6.1 Perfect gas with constant specific heat coefficients.

For this model the energy fundamental relation $u = u(s, v)$ reads

$$u(s,v) = u_0 \left(\frac{v}{v_0}\right)^{-\frac{R}{c_v}} \exp\left[\frac{s-s_0}{c_v}\right] \tag{6.1}$$

where:

- $R = \frac{R_0}{m}$ is the gas constant, with $m$ the gas molar mass and $R_0 = 1.98$ cal/mole °K the universal constant,
- $c_v$ is the constant volume specific heat coefficient;
- Subscript $(\ )_0$ denotes values in a referential state and $u_0 = u(s_0, v_0)$

Equivalently, this model can be characterized by means of the following two independent equations of state:

$$\begin{aligned} pv &= RT \qquad (R = \text{const}) \\ u &= u_r + c_v T \quad (c_v = \text{const}) \end{aligned} \tag{6.2}$$

The expressions for the specific heat coefficients and the characteristic speeds can be readily obtained from equation (6.2) once their definitions has been taken into account.

Alternatively, logarithmic differentiation of eqs. (6.1) yields :

$$\frac{dT}{T} = \frac{ds}{c_v} - \frac{R}{c_v}\frac{dv}{v} \tag{6.3a}$$

$$\frac{dp}{p} = \frac{ds}{c_v} - \frac{R + c_v}{c_v}\frac{dv}{v} \tag{6.3b}$$

where account has been taken of the Gibbs relation which, upon equations (6.3), reads:

$$c_v\,dT = T\,ds - \frac{RT}{v}\,dv .$$

From equations (6.3) one obtains:

$$\begin{aligned} c_p &= R + c_v = \text{const} \\ \gamma &= \frac{c_p}{c_v} = 1 + \frac{R}{c_v} = \text{const} \end{aligned} \tag{6.4a}$$

and:

$$\begin{aligned} a_L^2 &= -v^2\left(\frac{\partial p}{\partial v}\right)_s = \gamma RT \\ a_N^2 &= -v^2\left(\frac{\partial p}{\partial v}\right)_T = RT \end{aligned} \tag{6.4b}$$

and one recovers, as necessary, that

$$\frac{a_L^2}{a_N^2} = \gamma$$

Notice that, for this model:

- For a given gas, the characteristic speeds depend only upon the absolute temperature.
  Their square are inversely proportional to the molar mass.
- The equality (6.4) becomes :

$$c_p - c_v = R$$

- The specific heat ratio is constant and, upon stability requirements, is always greater than one.
- In terms of the molar specific heat $C_v = m\, c_v$ eq. (6.4) becomes:

$$\gamma = 1 + \frac{R_0}{C_v}$$

  where the number $\left(R_0 / C_v\right)$ depends on the type of the gas molecules.
- According to what established by means of the kinetic theory of gasses, this model rests on the following simplifying hypothesis.
  1) the volume occupied by the molecules themselves is negligible compared to that available to the gas
  2) molecular interaction is negligible
  3) the internal degrees of freedom of the molecule (rotational, vibrational, electronic) are either excited

"classically" (i.e. to each of them there corresponds an amount of energy, per mole, equal to $R_0T/2$) or unexcited.

The total number of rotational $(n_r)$ and vibrational $(n_v)$ degrees of freedom of a molecule depends upon the number $(n)$ of the atoms in the molecule and their configuration. In particular: for monoatomic molecules $n_r = n_v = 0$ for polioatomic linear molecules $n_r=2, n_v=3n-5$; for non linear polioatomic molecules $n_r=3$; $n_v=3n-6$. The subject model can be applied to real gases (and, in particular, to air) in the pressure and temperature ranges for which the above mentioned hypothesis are valid.

The first two hypothesis can be accepted provided $p \ll p_c$ and $T \gg T_c$ where $p_c$ and $T_c$ are the critical values of pressure and temperature. The third hypothesis can be accepted when the absolute temperature of the gas is sufficiently different from the vibrational characteristic temperatures of the molecules. Let $T_{v_i}$ be the characteristic temperature for the $i$-th vibrational degree of freedom of the molecule. For $T \ll T_{v_i}$, the $i$-th vibrational degree of freedom is not excited (i.e. does not contribute to the specific heats); for $T \gg T_{v_i}$ it is excited classically (i.e. it gives a constant contribution to $C_v$ equal to $\frac{R_0}{2}$). When $T$ is of the same order as $T_{v_i}$ the corresponding degree of freedom is excited but not "classically". Its contribution to $c_v$ is no longer constant but depends upon the absolute temperature $T$. (For all gases at normal temperature the rotational degrees of freedom can be considered to be excited classically. Their contribution to $c_v$ is therefore equal to $R_0$ for linear molecules and $3/2\,R_0$ for non linear molecules).

The average critical values for air are : $T_c = 132,4°K$ and $p_c = 37,2$ atm. Considering only its main constituents ($O_2$ and $N_2$) each of them has only one vibrational degree of freedom and the corresponding characteristic vibrational temperatures are $T_{v\,O_2} = 2240°K$; $T_{v\,N_2} = 3350°K$. The contributions due to the electronic degrees of freedom begin to be appreciable when $T$ is of the order of 10000°K . For most practical purposes air can be described by the perfect gas, constant specific heat model within a temperature range from ambient up to $6 \div 700°K$ and a pressure range from ambient up to few tens of atmospheres. More specifically, the model gas is biatomic with the vibrational (and, a fortiori, the electronic) degrees of freedom unexcited and the two rotational degrees of freedom excited classically.

Such a gas will have then 3 (transitional) + 2 (rotational) = 5 classically excited degrees of freedom. Hence, for air, within the above mentioned temperature and pressure ranges:

$$\frac{c_v}{R_0} = \frac{5}{2}\;;\quad \frac{c_p}{R_0} = \frac{7}{2}\;;\quad \gamma = 1.4$$

For the enthalpy one has:

$$h = u + pv = c_p T$$

Alternate equivalent and useful expressions for the enthalpy are:

$$h = \frac{c_p}{R}\frac{p}{\varrho} = \frac{\gamma}{\gamma-1}\frac{p}{\varrho} = \frac{a_L^2}{\gamma-1}$$

In terms of non-dimensional quantities

$$\hat{u} = \frac{u}{u_0 - u_2} \; ; \quad \hat{v} = \frac{v}{v_0} \; ; \quad \hat{s} = \frac{s - s_0}{c_v}$$

$$\hat{p} = \frac{v_0 p}{u_0} \; ; \quad \hat{T} = \frac{c_v T}{u_0}$$

the energy fundamental relation and the two state equations become:

$$\hat{u} = (\hat{v})^{1-\gamma} \exp(\hat{s})$$

$$\hat{p}\hat{v} = (\gamma - 1)\hat{T}$$

$$\hat{u} = \hat{T}$$

Hence, non-dimensionally, a particular perfect, constant specific heat gas is uniquely and completely characterized by a particular value of the number $\gamma$.

To conclude, we shall compute, for this model, the isentropic second derivative of the pressure with respect to the specific volume (or, equivalently, the third derivative $u_{vvv}$ of the energy fundamental function). This quantity is not affected by the stability rules and can thus have any sign. As it will be shown later, the sign of this derivative plays an essential rule in fluid dynamics.

From the relation

$$\left(\frac{\partial p}{\partial v}\right)_s = -\gamma \frac{p}{v}$$

one readily gets

$$\left(\frac{\partial^2 p}{\partial v^2}\right)_s = \frac{p}{v^2}\gamma(\gamma+1)$$

Hence, in this model, this derivative is always positive.

In a similar manner one obtains:

$$\left(\frac{\partial a_s^2}{\partial \varrho}\right) = -v^2\left(\frac{\partial a_s^2}{\partial v}\right)_s = \gamma p v^2(\gamma-1) \qquad (6.5)$$

The isentropic rate of change of $a_s^2$ with the density depends on the isentropic second derivative of pressure with respect to the specific volume. Hence also the sign of $a_s^2$ is not affected by stability rule and depends, on the particular model chosen and in general, also on the state of the fluid.

Since $\gamma > 1$ upon stability, it follows from eq. (6.4) that, in this model, the Laplacian characteristic speed is an increasing function of the density at $s$ constant.

This particular feature of the perfect gas model plays an essential role in the establishment fields.

6.2. Systems with internal rate processes.

Statistical thermodynamics provide a number of models for "simple" systems and, in particular, for simple gases. When dealing with a gas mixture, the knowledge of the fundamental relations of each "pure" gas is not sufficient to determine the fundamental relation of the mixture and additional assumptions are needed. A rather widely applicable one is that leading to the so called "ideal mixture" model which, for a system in thermal equilibrium, implies the two following hypothesis: a) each gas occupies the entire volume of the mixture; b) any fundamental relation of each gas in the mixture has the same form as the corresponding fundamental relation for the "pure" gas.

When the mixture is not in thermal equilibrium each gas, as seen, is itlself a composite system. The fundamental relation for such a pure gas is readily obtained, to within the shifting equilibrium hypothesis, from those of its subsystems upon the additivity property of extensive quantities. To obtain the fundamental relation for the mixture we shall extend to the non equilibrium case the above mentioned ideal mixture assumption. Many of the relations valid for ideal mixtures in thermal equilibrium are directly transferable to the non-equilibrium case provided the generalized potentials $\underset{\sim}{h}$ and $\underset{\sim}{f}$ are substituted for the "classical" potentials $h$ and $f$. Thus, for instance, the partial specific quantities of the generalized enthalpy are equal to the generalized enthalpies of the "pure" gases, i.e.:

$$\hat{\underset{\sim}{H}}_i = \underset{\sim}{h}_i = h_i - \sum_{j=1}^{k_i} \left[\Theta_j^{(i)} - T\right] s_j^{(i)} \tag{6.5}$$

where $h_i$ is the enthalpy of the pure gas and the $s_j^{(i)}$ are the entropies of its $(k_i)$ vibrational degrees of freedom which are not in mutual thermal equilibrium.

In what follows we shall furnish the fundamental relation in terms of either the classical or generalized Helmotz potential, the corresponding equations of state and the expression for the internal energy and the classical and generalized enthalpy for three particular thermodynamic models which, although pertaining to rather simple systems, have proved to be very useful tools in the study of non-equilibrium flows.

1. Ideal biatomic vibrating gas

In this model, the biatomic vibrating gas is considered as a composite system formed by the following sub-systems:

1) The ensemble of the translational and rotational degrees of freedom of the molecule
2) The ensemble of the vibrational degree of the

molecule.

Upon the shifting equilibrium hypothesis, we can formulate a fundamental relation for the composite system even when the two subsystems are not in thermal equilibrium.

The subject model is based upon the following hypothesis:

i) the specific degrees of freedom of the subsystems 1) and 2) and of the composite system, are, respectively, two, one and three:

ii) the generalized specific Helmotz potential $\underset{\sim}{f}$ for the composite system (defined as the second order Legendre transform of the internal energy with respect to the specific entropies $s^{(1)}$ and $s^{(2)}$ of the two subsystems).

Three different models will be assumed to describe the subsystem (2): classical excitation b) Lighthill model (i.e. vibrational degree of freedom excited to halt its classical values) ; c) one-dimensional harmonic oscillator.

If we denote by $T$ the absolute temperature of subsystem (1) and by $\Theta$ that of subsystem (2), the natural variables of $\underset{\sim}{f}$ are $(T, v, \Theta)$ and the model is characterized by the following fundamental relation:

$$\underset{\sim}{f}(T, v, \Theta) = \frac{R_0 \Theta}{M_2} \left\{ \begin{array}{c} \ln(\bar{\Theta}/\Theta) \\ \frac{1}{2}\ln(\bar{\Theta}/\Theta) \\ \ln\left[1-\exp(-\bar{\Theta}/\Theta)\right] \end{array} \right\} + \begin{array}{l} \text{a) (classical)} \\ \text{b) (Lighthill)} \\ \text{c) (harm. osc.)} \end{array} \tag{6.6}$$

$$+ f_{2r} + (f_{2o} - f_{2r}) \frac{T}{T_o} - \frac{R_0 T}{m_2} \left[ \ln\left(\frac{v}{v_o}\right) + \frac{5}{2} \ln\left(\frac{T}{T_o}\right) \right] \tag{6.6}$$

where: $R_0$ is the universal constant, $m_2$ the molar mass of the biatomic gas, the subscript $(o)$ denotes values in a fiducial state, $f_{2r}$ is a reference value for $\underset{\sim}{f}$ and the quantities $\bar{\Theta}$ and $(f_{2o} - f_{2r})$ are constants related to the structure of the molecule. Specifically, $\bar{\Theta}$ is the vibration characteristic temperature

$$\bar{\Theta} = \frac{h\nu}{K}$$

where $h$ is the Planck's constant, $K$ the Boltzman's constant and $\nu$ the natural vibration frequency of the molecule $\left[\bar{\Theta} = 2240\,°K \text{ for } O_2 \text{ and } \bar{\Theta} = 3350°\,K \text{ for } N_2\right]$ and :

$$f_{2o} - f_{2r} = \frac{R_0 T_o}{m_2} \left\{ \ln\left(\frac{T}{T_o}\right) - (1 + \sigma_{o2}) \right\}$$

$$\sigma_{o2} = \ln \left[ \frac{(2\pi m_2 K T_o)^{3/2} m_2 v_o \varepsilon_{o2}}{h^3} \right] \tag{6.7}$$

$$T = \frac{\eta h^2}{8\pi^2 K I}$$

where $\varepsilon_{o2}$ is the degree of degeneracy of the ground state. $\bar{T}$ is the characteristic temperature for rotation, $\eta$ is equal to two or one according to whether or not molecule is symmetric, and $I$ is the polar moment of the molecule.

The harmonic-oscillator model reduces to the classical excitation model for $(\bar{\Theta}/\Theta) \ll 1$.

The Gibbs relation reads:

$$d\underset{\sim}{f} = -s^{(1)} dT - p\, dv - s^{(2)} d(\Theta - T)$$

so that from the fundamental relation one obtains the following three independent equations of state:

$$s^{(1)} = -\frac{\partial \underset{\sim}{f}}{\partial T} = \frac{R_0}{m_2}\left\{ s_{o2} + \frac{5}{2}\ln\left(\frac{T}{T_0}\right) + \ln\left(\frac{v}{v_0}\right)\right\}$$

$$p = -\frac{\partial \underset{\sim}{f}}{\partial T} = \frac{R_0}{m_2}\,\frac{T}{v}$$

$$s^{(2)} = s_v = -\frac{\partial f}{\partial \Theta} = \frac{R_0}{m_2}\begin{cases} 1 + \ln(\bar{\Theta}/\Theta) & \text{(a)} \\ \frac{1}{2}\left[1 + \ln(\bar{\Theta}/\Theta)\right] & \text{(b)} \\ \dfrac{\Theta/\Theta}{\exp\left(\frac{\bar{\Theta}}{\Theta}\right)} - \ln\left[1 - \exp(-\bar{\Theta}/\Theta)\right] & \text{(c)} \end{cases} \tag{6.8}$$

with:

$$s_{o2} = \frac{5}{2} + \frac{f_{o2} - f_{or}}{T_0}$$

The total entropy of the gas is

$$s = s^{(1)} + s^{(2)} = s^{(1)} + s_v$$

The specific internal energy $(u)$ the enthalpy $(h)$ can be obtained from equations (6.6) and (6.8) on account of their definitions:

$$u = \underset{\sim}{f} + Ts^{(1)} + \Theta s^{(2)} = \underset{\sim}{f} + Ts + (\Theta - T) s_v$$

$$h = u + pv$$

[notice that the well known relation $u = -T^2 \dfrac{\partial (f/T)}{\partial T}$ is not valid for systems not in thermal equilibrium]. One gets:

$$u = u_{2r} + \frac{5}{2}\frac{R_0 T}{m_2} + u^{(2)}(\Theta)$$

$$h = u_{2r} + \frac{7}{2}\frac{R_0 T}{m_2} + u^{(2)}(\Theta)$$

where:

$$u^2(\Theta) = \frac{R_0}{m_2} \begin{cases} \Theta & \text{a)} \\ \dfrac{\Theta}{2} & \text{b)} \\ \dfrac{\bar{\Theta}}{\exp(\bar{\Theta}\ \Theta)-1} & \text{c)} \end{cases}$$

For the model b) the generalized enthalpy $\underset{\sim}{h} = h - (\Theta - T)s^{(2)}$ is given by:

$$\underset{\sim}{h} = u_r + \frac{4 R_0 T}{m_2} + \frac{\Theta - T}{2} \ln(\bar{\Theta}/\Theta)$$

Notice the following special cases, corresponding to conditions of constrained and unconstrained equilibrium of the subject composite system:

A) Frozen energy exchanges between subsystems 1) and 2) (i.e. "unexcited" vibrational degree of freedom). In this case $s^{(2)}$ and, hence, $\Theta$ are constant. Their contribution in the fundamental relation and in the expression for the internal energy and enthalpy, can be included in the (arbitrary) reference values $(f_{2r}, u_{2r})$ and the third independent equation of state (6.8) is inessential. The composite system has only two degrees of freedom. This model usually assumed for air at moderate energy levels (perfect gas model with constant specific heats).

B) Equilibrium energy exchanges between subsystem (1) and (2) (i.e. biatomic vibrating gas in thermal equilibrium). In this case $\Theta = T$ the composite system has again only two specific degrees of freedom; the explicit form for $\underset{\sim}{f}, u$, depends on the model assumed for the vibrational degree of freedom. Models a) and b) lead to the perfect gas constant specific heat model. Model c) leads to the perfect model with temperature dependent specific heats. The third equation of state (6.8) yields the equilibrium value $\mathfrak{s}_c^{(2)}$ of the entropy of subsystem (2).

## 2. Ideal biatomic dissociating gas. Lighthill model.

In this composite system there are two principal subsystems formed by the atoms [subsystem (1)] and the molecules [subsystem (2)].

In general the subsystem (2) itself may be a composite system and, in addition, the two subsystems may not be in thermal equilibrium. This more general case will be considered later on. Here we deal with the Lighthill model which is based on the following assumptions:

i) the biatomic molecules [subsystem (2)] constitute an equilibrium system with the vibrational degree of freedom excited to half its classical value;

ii) the atoms [subsystem (1)] are in thermal equilibrium with the molecules.

iii) the shifting equilibrium hypothesis applies for the process of exchange of mass between the two subsystems;

iv) the ideal-mixture hypothesis holds and each subsystem behaves as a perfect gas.

Thus, if $c$ is the atoms mass concentration; the Helmotz potential for the Lighthill model is:

$$(6.9) \qquad f(T,v,c) = c\, f_1\left(T, \frac{v}{c}\right) + (1-c)\, f_2\left(T, \frac{v}{1-c}\right)$$

where the $f_i$ are the specific Helmotz potentials of the atoms $(i=1)$ and molecules $(i=2)$.

For $f_1$, one has, in strict analogy with eq. (6.6):

$$f_1 = f_1\left(T, \frac{v}{c}\right) = f_{1r} + \left(f_{10} - f_{1r}\right)\frac{T}{T_0} -$$

$$(6.10) \qquad - \frac{R_0 T}{m_1}\left[\ln\left(\frac{v}{c v_0}\right) + \frac{3}{2}\ln\left(\frac{T}{T_0}\right)\right]$$

where:

$$(6.11) \qquad f_{10} - f_{1r} = - \frac{R_0 T_0}{m_1}\left(1 + \sigma_{01}\right)$$

and $\sigma_{01}$ has the expression given by equation (6.7) with $m_1$ and $\varepsilon_{01}$ replacing $m_2$ and $\varepsilon_{02}$.

The expression for $f_2[T, v/(1-c)]$ is obtained from equation (6.6) model (b), with $\Theta = T$ and $v$ being replaced by $v/(1-c)$ [ideal-mixture hypothesis].

If one assumes the same fiducial state $(T_0, v_0)$ for the two gases, notice that $m_2 = 2m_1$ takes $f_{2r} = 0$ so that $f_{1r} = = \frac{D_0}{2m_1} = \frac{D_0}{m_2}$ where $D_0$ is the dissociation energy per mole, equation (6.6), (6.9) and (6.10) lead to:

$$f = \frac{D_0 c}{m_2} + \frac{R_0 T}{m_2}\left[3 - \frac{s_0 m_2}{R_0} - c\right] -$$

$$- \frac{R_0 T}{m_2} \ln\left[\left(\frac{T}{T_0}\right)^3 \left(\frac{v}{v_0}\right)^{1+c} c^{-2c}(1-c)^{-(1-c)}\right] \tag{6.12}$$

where it has been assumed that :

$$\frac{(f_{10} - f_{20} - f_{1r}) m_2}{R_0 T_0} + \frac{1}{2} \ln\left(\frac{T_0}{\Theta}\right) = -1 \tag{6.13}$$

and the constant $s_0$ is given by:

$$\frac{m_2 s_0}{R_0} = 3 - \frac{m_2 f_{20}}{R_0 T_0} + \frac{1}{2}\ln\left(\frac{T_0}{\Theta}\right) =$$

$$= (4 + \sigma_{02}) + \ln\left[\left(\frac{T_0}{T}\right)\left(\frac{T_0}{\Theta}\right)^{1/2}\right] \tag{6.14}$$

with the last equality following from eq. (6.7).

Eq. (6.13) amounts to choosing a particular value for $v_0$, related to the physical properties of the atom and of the molecule. Indeed, on account of eqs. (6.11) and (6.7), (6.13) becomes :

$$\sigma_{02} - 2\sigma_{01} + \ln\left[\left(\frac{T_0}{\bar{T}}\right)\left(\frac{T_0}{\bar{\Theta}}\right)^{1/2}\right] = 0$$

so that, upon equation (6.7) it gives:

$$\upsilon_0 = \frac{2h^3 \varepsilon_{02}}{\varepsilon_{01}^2 \bar{T} \bar{\Theta}^{1/2} m_1^{5/2} (\pi K)^{3/2}} \tag{6.15}$$

The Gibbs relation reads:

$$df = -s\,dT - p\,d\upsilon + A\,dc$$

where $A = (g_1 - g_2)$ is the affinity of the dissociation. The three independent equations of state read:

$$\frac{m_2(s-s_0)}{R_0} = c + \ln\left[\left(\frac{T}{T_0}\right)^3 \left(\frac{\upsilon}{\upsilon_0}\right)^{1+c} c^{-2}(1-c)^{c-1}\right]$$

$$p\upsilon = \frac{R_0 T}{m_2}(1+c) \tag{6.16}$$

$$\frac{m_2 A}{D_0} = 1 + \frac{R_0 T}{D_0} \ln\left[\left(\frac{\upsilon_0}{\upsilon}\right) \frac{c^2}{1-c}\right].$$

The expression for the internal energy is readily obtained from the definition $n = -T^2\left[\partial(f/T)/\partial T\right]_{c,\upsilon}$ and reads:

$$\frac{m_2 n}{D_0} = c + \frac{3R_0}{D_0} T \tag{6.17}$$

and for the enthalpy one has:

$$\frac{m_2 h}{D_0} = \left(1 + \frac{R_0 T}{D_0}\right) c + 4 \frac{R_0 T}{D_0} \tag{6.18}$$

The Lighthill model is a particular case of a more general model for an ideal single reacting gas mixture in thermal equilibrium [see Ref. 4] . It can be shown that the model proposed by Lighthill amounts to assume that the constant volume specific heat is an invariant with respect to the dissociation.

3. Ideal gas mixture with an arbitrary number of rate processes.

In the previous paragraphs two particular thermodynamic models have been given in each of which only one type process was present.

In a more general case one may consider that an arbitrary number of processes of either type (i.e. energy and mass exchange among the subsystem) may take place. Such a model can be found in Ref.(2). The basic hypothesis used in constructing it were mentioned at the beginning of this section and amount to assume that the generalized Helmotz potential of the mixture (defined as the Legendre transform of the internal energy with respect to the entropies of all subsystems), is given by the sum of the generalized Helmotz potentials of each "pure" gas occupying the entire volume of the mixture.

In the general case when there is an arbitrary

number of reactions, of internal and/or translational degrees of freedom not in thermal equilibrium the corresponding formulae are too lengthy to be reported here.

We shall limit ourselves to one of the two models proposed by Monti and Napolitano (Ref.2) namely that of an ideal dissociating and vibrating diatomic gas. Such a system has four specific degrees of freedom and it is constituted by the following subsystems:

1) Atomic gas: 2) Ensemble of the translational and rotational degrees of freedom of the molecules, 3)Ensemble of the vibrational degree of freedom of the molecules. Subsystems (1) and (2) are in thermal equilibrium with the other two and shall be described by the Lighthill model.

The set of natural variables for the generalized Helmotz potential is the set $(T, v, c, \Theta)$ (with the notations of the previous paragraphs) and one has:

$$\underset{\sim}{f}\left(T, v, c, \Theta\right) = c f_1\left(T, \frac{v}{c}\right) + (1-c)\underset{\sim}{f}_2\left(T, \frac{v}{1-c}, \Theta\right) \tag{6.19}$$

with $f_1$ given by equation (6.10) and $\underset{\sim}{f}_2$ given by equation (6.6) model (b) with $v$ replaced by $v/(1-c)$.

Proceeding as for the Lighthill model and with the same notations, one gets:

$$\underset{\sim}{f} = \frac{D_0 c}{m_2} + \frac{R_0 T}{m_2}\left[3 - \frac{m_2 s_0}{R_0} - c\right] - \frac{R_0 \Theta}{m_2}\left(\frac{1-c}{2}\right)\ln(\bar{\Theta}/\Theta) -$$

$$- \frac{R_0 T}{m_2}\ln\left[\left(\frac{T}{T_0}\right)^3\left(\frac{v}{v_0}\right)^{1+c}\left(\frac{T}{\Theta}\right)^{\frac{c-1}{2}} c^{-2c}(1-c)^{c-1}\right] \tag{6.20}$$

with $s_0$ and $v_0$ still given by eqs. (6.14) and (6.15).

From the Gibbs relation written as:

$$d\underset{\sim}{f} = -s\,dT - p\,dv + A\,dc - s_v\,d(\Theta - T)$$

one gets the four independent equations of state:

$$\frac{m_2(s - s_0)}{R_0} = c + \ln\left[\left(\frac{T}{T_0}\right)^3\left(\frac{v}{v_0}\right)^{1+c}\left(\frac{\Theta}{T}\right)^{\frac{1-c}{2}} c^{-2c}(1-c)^{c-1}\right]$$

$$pv = \frac{R_0 T}{m_2}(1+c) \tag{6.21}$$

$$\frac{m_2 A}{D_0} = 1 + \frac{R_0 T}{D_0}\ln\left[\frac{c^2}{1-c}\left(\frac{v_0}{v}\right)\left(\frac{\bar{\Theta}}{T}\right)^{1/2}\left(\frac{\Theta}{\bar{\Theta}}\right)^{\frac{\Theta}{2T}}\right]$$

$$\frac{m_2 s_v}{R_0} = \frac{1-c}{2}\left[1 + \ln(\Theta/\bar{\Theta})\right].$$

From these relations one can obtain all other thermodynamic quantities. In particular for the internal energy and the enthalpies one has:

$$\frac{m_2 n}{D_0} = c + \frac{3R_0T}{D_0} + \frac{1-c}{2}\frac{R_0}{D_0}(\Theta - T)$$

$$\frac{m_2 h}{D_0} = \left(1 + \frac{R_0T}{D_0}\right)c + \frac{4R_0T}{D_0} + \frac{1-c}{2}\frac{R_0}{D_0}(\Theta - T)$$

(6.22)

$$\underset{\sim}{h} = h - (\Theta - T)s_v = \frac{D_0}{m_2}\left\{\left(1 + \frac{R_0T}{D_0}\right)c + \frac{4R_0T}{D_0} + \right.$$

$$\left. + \frac{1-c}{2}(\Theta - T)\frac{R_0}{D_0}\ln(\bar{\Theta}/\Theta)\right\}.$$

This being the most general model we have presented here, all others must be recovered as particular cases of this one, corresponding to some of its possible different equilibrium conditions. Thus:

- the Lighthill model corresponds to states of a first order partial equilibrium, namely equilibrium with respect to the energy exchanges $(\Theta = T)$ with the fourth equation of state (6.21) giving the equilibrium value for the specific entropy of the vibrational degree of freedom;
- the ideal dissociating gas (model b) corresponds to states of

a first order constrained partial eqiulibrium, [the exchange of mass is 'frozen" at the value $c=0$]. Eq. (6.6) is recovered from eq.(6.20) with $c=0$, as readily checked. The third equation of state (6.21) yields the value of the affinity $A$ for these frozen states as function of the three remaining independent state parameters $(T, \Theta, v)$; and determines whether the constraint is active or not;

- the perfect gas model corresponds to a second order equilibrium, constrained with respect to the exchange of mass $(c=0)$ and either constrained (constant specific heat model, $\Theta$ = constant) or unconstrained (temperature dependent specific heat, $\Theta=T$) with respect to the exchange of energy.
- the full equilibrium states $(A=0, \Theta=T)$ describe a non-perfect gas model with two specific degrees of freedom. The third equation of state defines implicitly (with $\Theta=T$) the equilibrium value of the atom's concentration and, as before, the fourth equation gives, for $\Theta=T$ and $c=c_e$, the equilibrium value of the specific entropy of the vibrational degree of freedom of the molecule.

As repeatedly stated, macroscopic thermodynamic can only say what happens when certain types of equilibrium prevail. Which type of equilibrium actually occurs in a given particular case depends on a number of ratios of characteristic times which can be evaluated only when the rate processes have been characterized. The completely different thermodynamic behaviour of the above mentioned models will naturally lead to completely different flow field properties. Hence the need for a thorough and careful analysis of the types of equilibrium which prevail during a given evolution of the system.

## 4. Second model for the ideal biatomic vibrating and dissociating gas.

To account properly for the fact that the energy associated with the degree of vibration of the molecule cannot increase indefinitely because the molecules will eventually dissociate, another model for the ideal vibrating and dissociating gas has been proposed by Harnmerling, Teare and Kwel (ref. 6). The only difference between this model and the previous one consists in the fact that the vibrational degree of freedom of the molecules is described by the simple harmonic oscillator with cut-off at an energy equal to the dissociation energy (per mole) $D_0$.

Hence, in this new model, only the terms expressing the contributions due to the vibrational temperature ( $\Theta$ ) change. Specifically, the expression for $\underset{\sim}{f}_2(T, \Theta, \upsilon / 1-c)$ in eq. (6.19) is to be taken from eq. (6.6) model (c), with $\upsilon$ replaced by $\upsilon/1-c$ from which one subtracts the term:

$$\frac{R_0 \Theta}{2} \ln \left[ 1 - \exp\left( - \frac{D_0}{R_0 \bar{\Theta}} \right) \right]$$

which represents the "cut-off" for the allowed vibrational energy levels. The quantity $(D_0 / R_0)$ can be interpreted as a characteristic temperature for dissociation, $\bar{T}_D$. For $O_2$ and $N_2$ it is, respectively, $\bar{T}_D = 59{,}390°K$ and $\bar{T}_D = 180{,}500°K$. This correction is therefore usually very small. [The ratio $(\bar{T}_D / \bar{\Theta})$ for $O_2$ and $N_2$ is approximately equal to 26 and 54 respectively]. For values of $(\Theta / \bar{\Theta})$ up to 5 the corresponding contributions to the internal energy and the entropies are very small (Ref. 5). However, as we shall see, this model with cut-off plays an important part in the kinetics of the rate processes. For this reason we shall present here its relevant equations.

To obtain the generalized Helmotz potential we start again from eq. (6.9) but now we take for $\underset{\sim}{f}_2$ the expression:

$$\underset{\sim}{f}_2 = f_{2r} + (f_{2o} - f_{2r}) \frac{T}{T_0} - \frac{R_0 T}{m_2} \left\{ \ln \left[ \frac{v}{v_0(1-c)} \right] + \frac{5}{2} \ln \frac{T}{T_0} \right\} - \frac{R_0 \Theta}{m_2} \ln \left[ Q(\Theta) \right]$$

where:

$$Q(\Theta) = \frac{1 - \exp\left(-\dfrac{D_0}{R\Theta}\right)}{1 - \exp\left(-\dfrac{\bar{\Theta}}{\Theta}\right)} \tag{6.23}$$

is the "partition function" of the vibrational degree of freedom. We shall assume $f_{2r}=0$; $f_{1r}=\frac{D_0}{m_2}$ as before but instead of the relation (6.13) we shall impose the condition:

$$f_{20} = f_{10} - f_{1r} + \frac{R_0 T_0}{m_2} = f_{10} + \frac{R_0 T_0 - D_0}{m_2}$$

[i.e. we change the fiducial value for $v_0$]. We thus get:

$$f = \frac{c\,D_0}{m_2} + \frac{R_0 T}{m_2}\left\{\frac{m_2 f_{10}}{R_0 T_0} + \left(1 - \frac{D_0}{RT_0}\right) - c\right\} -$$

$$- \frac{R_0 T}{m_2}\,\ell n\left[\left(\frac{v}{v_0}\right)^{1+c}\left(\frac{T}{T_0}\right)^{\frac{5+c}{2}} c^{-2c}(1-c)^{c-1}\right] +$$

$$+ \frac{R_0 \Theta}{m_2}(c-1)\,\ell n\left[Q(\Theta)\right] \tag{6.24}$$

The corresponding equations of state can be obtained readily and read:

$$s = -\frac{\partial f}{\partial T} = \frac{R_0}{m_2}\left[\left(\frac{5}{2} - \beta_0\right) + \left(\frac{1}{2} - \beta\right)c\right] +$$

$$+ \frac{R_0}{m_2}\,\ell n\left\{\left(\frac{v}{v_0}\right)^{1+c}\left(\frac{T}{T_0}\right)^{\frac{5+c}{2}} c^{-2c}(1-c)^{c-1}\right\} +$$

$$+ \frac{R_0}{m_2}(1-c)\quad \ell n\, Q + \frac{\Theta\, Q'}{Q} \tag{6.25}$$

$$p\upsilon = \frac{R_0 T}{m_2}(1+c)$$

$$s_\upsilon = -\frac{\partial f}{\partial \Theta} = \frac{R_0}{m_2}(1-c)\left[\ln Q(\Theta) + \frac{\Theta Q'}{Q}\right]$$

$$A = \frac{\partial f}{\partial c} = \frac{D_0}{m_2} + \frac{R_0 T}{m_2}(\beta + 1) + R_2 \Theta \ln Q(\Theta) +$$

$$+ \frac{R_0 T}{m_2} \ln\left\{\left(\frac{T_0}{T}\right)^{\frac{1}{2}}\left(\frac{\upsilon}{\upsilon_0}\right)\frac{c^2}{(1-c)}\right\}$$

with:

$$\beta_0 = \frac{f_{20} m_2}{R_0 T_0} \quad ; \quad \beta = \ln\left(\frac{T_0}{T}\right) - \left[1 + 2\sigma_{01} - \sigma_{02}\right]$$

$$Q'(\Theta) = \frac{d\,Q(\Theta)}{d\,\Theta}$$

$$Q(\Theta) = \begin{cases} \Theta / \bar{\Theta} \\ (\Theta/\bar{\Theta})^{1/2} \\ \left[1 - \exp(-\bar{\Theta}/\Theta)\right]^{-1} \\ \dfrac{1 - \exp(-D_0/R_0\Theta)}{1 - \exp(-\bar{\Theta}/\Theta)} \end{cases}$$

## 7. BALANCE EQUATIONS

The total number of field unknowns for any fluidynamic problem is given by the sum of those required to characterize the "local" thermodynamic state of the medium plus those required to describe its "kinematical" state. As seen, a state of a thermodynamic system with $(d+1)=(n+k+2)$ extensive degrees of freedom is characterized by a set of $(d+1)$ independent extensive quantities. We shall assume that the kinematical state of the mixture is sufficiently characterized by its local momentum (so called single-fluid theory), which is also an extensive property. Hence the fluid-dynamic field is characterized by an independent set of $(d+2)$ extensive quantities or, per unit mass, by a set of $(d+1)$ specific quantities (which we shall refer to as basic or "primitive" unknowns).

For any such extensive quantity $(X)$ one can formulate a balance equation expressing the obvious logical fact that the total value of $X$ in any given (fixed) system can vary only as a consequence of the interaction between the system and its environment and the eventual production of $X$ within the system itself. In this section we shall derive the balance equation for an arbitrary extensive quantity (scalar and/or vectorial) and discuss some of its relevant properties. The general relationships derived here will be later applied to the basic set of field variables.

## 7.1 Integral and differential forms of the balance equations.

Let $\Sigma$ be the system, $V$ its volume, $D$ its surfaces separating it from its environment $\Sigma'$ and suppose $V$, $D$ fixed with respect to a given Galileian frame $(G)$

Define the flux of $x$, $\underline{J}_x$, as the operator giving, in direction and intensity, the quantity of $x$ flowing per unit area and time, as measured in $G$. The flux $\underline{J}_x$ is vectorial (resp. tensorial) if $x$ is a scalar (resp. a vector, $\underline{x}$ ). Any vectorial flux can be expressed, formally, as:

$$\underline{J}_x = \varrho x \underline{W} \tag{7.1}$$

where, as usual, low case letters denote specific values referred to the unit mass, $\varrho$ is the total mass density and $\underline{W}$ is a velocity. To within the limits of validity of non relativistic mechanics, all thermodynamic quantities so far introduced are Galileian invariants.

The law of transformation of the flux of a scalar Galileian invariant quantity from one Galileian frame $(G)$ to another $(G')$ is the same as that of a velocity. The transformation law for a tensorial flux depends on the law of transformation of the flowing quantity

Define the production $\dot{x}$ of $x$ as the quantity of $x$ being created $(\dot{x}>0)$ or destroyed $(\dot{x}<0)$ per unit volume and time, as measured in $G$. The production of a quantity has the same tensorial order as the quantity. Productions of all the scalar quantities we shall be interested in are Galileian invariant. When $\dot{x}$ vanishes identically the quantity $x$ will be called <u>conservative</u>.

When the convention that the positive unit vector $\underline{n}$ normal to the bounding surfaces $D$ of $\Sigma$ is oriented toward $\Sigma$ (i.e. outside of $\Sigma$) the integral forms of the balance equations for an arbitrary scalar ($x$) and vectorial ($\underline{x}$) quantity read:

$$\int_{v}\left(\frac{\partial \varrho x}{\partial t}\right) dv + \int_{D}\left(\underline{n}\cdot\underline{J}_x\right) dD = \int_{v} \dot{x}\, dv$$

(7.2)

$$\int_{v}\left(\frac{\partial \varrho \underline{x}}{\partial t}\right) dv + \int_{D}\left(\underline{n}\cdot\tilde{\underline{\underline{J}}}_x\right) dD = \int_{v} \dot{\underline{x}}\, dv$$

where, $\tilde{\underline{\underline{J}}}_x$ is the transpose of $\underline{\underline{J}}_x$.
Upon the Gauss theorem, these equations can also be written as:

$$\int_{v}\left[\frac{\partial \varrho x}{\partial t} + \underline{\nabla}\cdot\underline{J}_x - \dot{x}\right] dv = 0$$

(7.3)

$$\int_{v}\left[\frac{\partial \varrho \underline{x}}{\partial t} + \underline{\nabla}\cdot\tilde{\underline{\underline{J}}}_x - \dot{\underline{x}}\right] dv = 0$$

Equations (7.3) are valid for any $v$. Hence, when the integrands are continuous, one obtains the following differential form of the balance equations:

$$\frac{\partial \varrho x}{\partial t} + \underline{\nabla} \cdot \underline{J}_x = \dot{x}$$

$$\frac{\partial \varrho \underline{x}}{\partial t} + \underline{\nabla} \cdot \underline{\underline{\tilde{J}}}_x = \dot{\underline{x}} \qquad (7.4)$$

For a non-relativistic mechanics, these equations are Galileian invariant.

To understand the true meaning of a balance equation, the following remarks are appropriate.

- Any time the balance equation for a given quantity is formulated, one introduces two additional unknowns: the flux and the production of the quantity itself. To distinguish them from the "primitive" field unknowns we shall refer to them as "induced" unknowns.
- The purely formal statement implied in the balance equations will be given physical substance when a number of "conservation postulates" will be imposed. These postulates will eliminate a number of productions. The remaining productions and fluxes will have to be characterized through suitable phenomenological relations.
- For a non-conservative quantity (i.e. for a quantity for which $\dot{x}$ does not vanish identically) there may be a degree of arbitrariness in the definition of the flux and of the production, only the quantity $(\dot{x} - \underline{\nabla} \cdot \underline{J}_x)$ being uniquely defined.

Hence, in general, a vectorial flux is defined unless an arbitrary vector $\underline{u}$ and, correspondingly, the production is defined unless the divergence of $\underline{u}$ . Similar considerations apply to the flux and the production of a vectorial quantity.

- For a conservative quantity the production $\dot{x}$ vanishes identically, as said. The corresponding balance equation will be refered to as a conservation equation. Their two possible forms read, for a scalar property:

$$\int_v \left(\frac{\partial \varrho x}{\partial t}\right) dv + \int_D (n \cdot \underline{J}_x)\, dD = 0$$

(7.5)

$$\frac{\partial \varrho x}{\partial t} + \underline{\nabla} \cdot \underline{J}_x = 0$$

The corresponding forms for vectorial quantities are obvious.

- Strictly speaking even for a conservative quantity the flux is not uniquely defined. However it results defined unless an inessential solenoidal vector (or tensor, if the quantity is vectorial).

7.2 Jump conditions.

When the integrands of equations (7.3) are not continuous there are discontinuities within the volume V or on its boundaries. These discontinuities may only be isolated, i.e. they are surfaces, lines or points. We are essentially interested in the first ones. The imposition of the balance equation across discontinuity surfaces leads to the so called "jump conditions" which relate the values of the flow properties on the two sides of the discontinuity surfaces.

We shall suppose that the "local" time derivatives $\left(\frac{\partial \varrho x}{\partial t}\right)$, the tensorial and vectorial fluxes and the vectorial productions $\dot{\underline{x}}$ remain finite whereas the scalar productions $\dot{x}$ may attain infinitely large values in such a way that one can define a finite production per unit surface and time $\dot{x}_s$.

Let $\sigma$ be a discontinuity surface which is fixed with respect to a certain Galileian frame G .

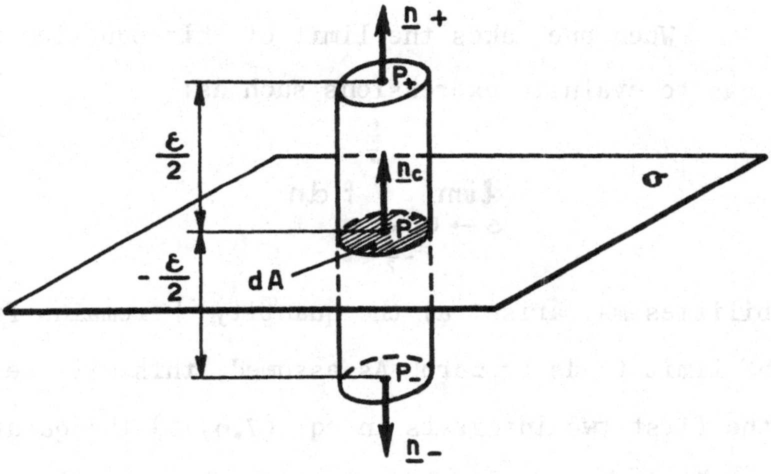


Consider a point P on $\sigma$ . Let $\underline{n}_c$ be the unit normal to $\sigma$ in P (which arbitrarily defines the "positive" side of the discontinuity surface), $dA$ an infinitesimal area on $\sigma$ around P, and C a cylinder, with normal cross-section $dA$ which is extended by $\varepsilon/2$ on the two sides of the surface, in a direction parallel to $\underline{n}_c$. On applying the balance equation (7.2) to this cylinder one gets:

$$dA\int_{-\frac{\varepsilon}{2}}^{\frac{\varepsilon}{2}}\left(\frac{\partial \varrho x}{\partial t}\right)dn+\int_{L_c}(\underline{n}\cdot\underline{J}_x)dD+\left[(\underline{n}_+\cdot\underline{J}_x)_{P_+}+\right.$$

$$(7.6) \qquad \left.+(\underline{n}_-\cdot\underline{J}_x)_{P_-}\right]dA = dA\int_{-\frac{\varepsilon}{2}}^{\frac{\varepsilon}{2}}\dot{x}\,dn$$

where the subscripts $\underline{P}_+$ and $\underline{P}_-$ denote the points where the quantities have to be calculated and $L_c$ is the lateral surface of the cylinder.

When one takes the limit of this equation for $\varepsilon\rightarrow 0$ one has to evaluate expressions such as:

$$(7.7) \qquad \lim_{\varepsilon\rightarrow 0}\int_{-\frac{\varepsilon}{2}}^{\frac{\varepsilon}{2}} f\,dn$$

Two possibilities may arise: a) the quantity $f$ remains finite so that the limit tends to zero. As assumed, this will be the case for the first two integrals in eq. (7.6) b) the quantity $f$ may attain infinitely large values in such a manner that :

$$\lim_{\varepsilon\rightarrow 0}\int_{-\frac{\varepsilon}{2}}^{\frac{\varepsilon}{2}} f\,dn = \text{finite}$$

In practice such a case may arise only in connection with some scalar production: the finite limit can then be interpreted as a "surface source" $\dot{x}_s$ defined as:

$$\lim_{\varepsilon \to 0} \int_{-\frac{\varepsilon}{2}}^{\frac{\varepsilon}{2}} \dot{x}\, dn = \dot{x}_s(P)$$

The "surface source" $\dot{x}_s$ is obviously function, in general, of the points $P$ of the discontinuity surface $\sigma$.

By noticing that as $\varepsilon \to 0$, $\underline{n}_+ \to \underline{n}_c$ and $\underline{n}_- \to \underline{n}_c$ and by denoting by $\delta f = (f_+ - f_-)$ the "jump' of the quantity $f$ across the discontinuity surface (i.e. the value $f_+$ of $f$ computed on the positive side of the surface minus the value of $f$ computed on the negative side) equation (7.6) becomes, for $\varepsilon \to 0$ :

$$\delta\left(\underline{n} \cdot \underline{J}_x\right) = \dot{x}_s$$

This is the required "jump form" of the balance equation for the scalar quantity $x$ . In particular, when $x$ is conservative one has, simply:

$$\delta\left[\underline{n} \cdot \underline{J}_x\right] = 0$$

i.e. the normal component of the flux is certainly continuous across $\sigma$.

For a vectorial quantity $x$ the jump condition reads:

$$\delta\left[\underline{n} \cdot \tilde{\underline{\underline{J}}}_x\right] = 0$$

since, for the cases of interest in this course, we shall only consider finite vectorial productions.

To summarize:

" If $\sigma$ is a discontinuity surface fixed with respect to a certain Galileian frame G , then the component $\underline{n} \cdot \underline{J}_x$ normal to $\sigma$ of the flux $\underline{J}_x$ (measured in G ) can be discontinuous across $\sigma$ if and only if the production of $x$ is not finite".

The jump conditions are used either to investigate the nature and properties of possible discontinuity surfaces that may exist in the fluid or to obtain "boundary conditions" that must be satisfied by the field variables at the solid surfaces bounding the fluid (or, more generally, at interfaces).

## 8. BALANCE EQUATIONS FOR THERMODYNAMIC STATE VARIABLES

Each of the $(k+n+2)$ thermodynamic state variables

$$\left[ S ; V ; m ; m_i , \ (1 \leq i \leq n-1) ; \ S_j , (1 \leq j \leq k) \right]$$

satisfies a balance equation. We shall deal here explicitly only with the forms given by eq. (7.3). A general discussion to the jump relations in connection with the existence of discontinuity surfaces will be given later.

## 8.1 Mass conservation - Continuity equation

Let $M$ be the total mass of the system. According to one of the basic postulate of non-relativistic mechanics its production vanishes identically. Hence if $\underline{J}_m$ is the mass flux in a Galileian frame $G$ , the differential form of the conservation equation for $M$ reads (in $G$ ):

$$\frac{\partial \varrho}{\partial t} + \underline{\nabla} \cdot \underline{J}_m = 0 \tag{8.1}$$

In this equation $\underline{J}_m$ is uniquely defined and, in turn, it defines the mass velocity $\underline{V}$ (in $G$ ) according to the relation:

$$\underline{V} = \frac{1}{\varrho} \underline{J}_m \tag{8.2}$$

In terms of the mass velocity $V$ , eq. (8.1) becomes:

$$\frac{\partial \varrho}{\partial t} + \underline{\nabla} \cdot (\varrho \underline{V}) = 0 \tag{8.3}$$

This equation is usually refered to as the continuity equation. It is Galileian invariant, i.e. in a new Galileian frame $G'$ which moves with respect to $G$ at a constant (linear) velocity $C$ , the continuity equation reads:

$$\frac{\partial \varrho}{\partial t'} + \underline{\nabla} \cdot (\varrho \underline{V}') = 0$$

where $\underline{V}' = \underline{V} - \underline{C}$ is the mass velocity in $G'$ and

(8.4) $$\frac{\partial}{\partial t'} = \frac{\partial}{\partial t} + \underline{C}\cdot\underline{\nabla}$$

If the motion is steady in a given Galileian frame, the corresponding mass flux $\underline{J}_m = \varrho\underline{V}$ is solenoidal.

If the density variations can be neglected (i.e. if the fluid behaves as an incompressible fluid) the mass velocity $\underline{V}$ is solenoidal.

By introducing the 'material" time derivative $(D/Dt)$ defined by:

(8.5) $$\frac{D}{Dt} = \frac{\partial}{\partial t} + \underline{V}\cdot\underline{\nabla}$$

the continuity equation can also be written as:

(8.6) $$\frac{1}{\varrho}\frac{D\varrho}{Dt} = -\underline{\nabla}\cdot\underline{V}$$

The "material" derivative is a time derivative evaluated by keeping constant the space coordinates characterizing suitably chosen initial configuration of the medium (i.e. it gives the time rates of change of properties of a given particle). Upon this interpretation or, formally, from equations (8.4) and (8.5) it follows that the operator $D/Dt$ is Galileian invariant.

$$\left[\text{i.e. } \frac{D}{Dt'} = \frac{\partial}{\partial t'} + \underline{V}'\cdot\underline{\nabla} = \frac{D}{Dt}\right]$$

## 8.2 Convective and diffusive fluxes

The total fluxes $\underline{J}_x$, as seen, are not Galileian invariant. It is clearly convenient, to single out from them a Galileian invariant part.

For a scalar Galileian invariant property $x$ this is readily done by letting:

$$\underline{J}_x = \varrho x \underline{V} + \underline{j}_x \qquad (8.7)$$

The first part $(\varrho x \underline{V})$ is called "convective flux". The second part $(\underline{j}_x)$ is called "diffusive flux". Both the total $(\underline{J}_x)$ and convective fluxes $(\varrho x \underline{V})$ depend on the Galileian frame G in which they are measured. On the contrary the diffusive flux is Galileian invariant and, therefore, it is to be considered an "intrinsic" "property" of the fluid.

The Galileian invariance of $\underline{j}_x$ follows immediately from that of $x$ and from eq. (8.7). Indeed, one can always let $\underline{J}_x = \varrho x \underline{W}_x$ (this relation in fact, defines the velocity $\underline{W}_x$) so that $\underline{j}_x = \varrho x (\underline{W}_x - \underline{V})$ and its invariance follows therefore from that of $x$.

From another point of view, $\underline{j}_x$ can also be considered as the flux measured in a frame (in general non-Galileian) moving with the velocity $\underline{V}$, i.e. with the velocity of the center of gravity of the considered particle. From these remarks it follows that when the balance equations are expressed in terms of diffusive fluxes the corresponding time rate of change will have to be expressed through the 'material' derivative.

Indeed, if eq. (8.7) is substituted in eq.(7.5) and account is taken of the continuity equation the balance equation will become:

$$\varrho \frac{Dx}{Dt} + \underline{\nabla} \cdot \underline{j}_x = \dot{x} \tag{8.8}$$

to the other form, in this form of the balance equation <u>each term</u> is Galileian invariant.

Stating it differently, when the scalar balance equation is written in the form (8.8) we should no longer worry about the particular Galileian frame we are using. The reduction of the vectorial balance equation to a similar, individually invariant form will be taken up later.

The previously mentioned arbitrariness in the definition of the total flux is transferred now into an arbitrariness in the definition of the diffusive flux. Whenever possible, we shall dispose of this arbitrariness by requiring that the diffusive fluxes be linear operators.

Notice that:

- By definition, the diffusive flux of total mass is zero $\left(\underline{j}_m = 0\right)$
- Quite obviously, eq. (8.8) reduces to an identity when applied

to the total mass (since $x = 1$, $\underline{j}_x = \dot{x} = 0$)

- When applied to the volume $V$ eq. (8.8) gives:

$$\varrho \frac{Dv}{Dt} = \dot{v} - \underline{\nabla} \cdot \underline{j}_v = \underline{\nabla} \cdot \underline{V} \qquad (8.9)$$

since $v = \frac{1}{\varrho}$. Upon the requirement that diffusive fluxes be linear operators it is $\underline{j}_v \equiv 0$ (for, from the relation $m = \varrho v$ it follows that $\underline{j}_v = \frac{1}{\varrho} \underline{j}_m$) so that $\underline{\nabla} \cdot \underline{V}$ is to be interpreted as the "production" of the volume (per unit volume and time)

- In terms of diffusive fluxes the jump relation reads:

$$\delta \left[ \underline{n} \cdot \left( \varrho x \underline{V} + \underline{j}_x \right) \right] = 0 \qquad (8.10)$$

8.3 Balance equations for thermodynamic state variables

Listing, for simplicity, only their differential form, individually invariant, the subject balance equations read with an obvious meaning of symbols:

$$\begin{aligned} \varrho \frac{Ds}{Dt} + \underline{\nabla} \cdot \underline{j}_s &= \dot{s} \\ \varrho \frac{Dv}{Dt} &= \underline{\nabla} \cdot \underline{V} \\ \varrho \frac{DC_i}{Dt} + \underline{\nabla} \cdot \underline{j}_{C_i} &= \dot{C}_i \quad (1 \le i \le n-1) \\ \varrho \frac{Ds_j}{Dt} + \underline{\nabla} \cdot \underline{j}_{s_j} &= \dot{s}_j \end{aligned} \qquad (8.11)$$

The balance equation for the n-th constituent of the mixture is an identity on account of the facts that $\sum_{i=1}^{n} C_i = 1$ (by definition); $\sum_{i=1}^{n} \dot{C}_i = 0$ (upon the conservation of mass); $\sum_{i=1}^{n} \underline{j}_{C_i} = 0$ (upon the definition of the mass velocity $\underline{V}$ )

## 8.4 Transformation of the mass balance equations for the constituents of the gas mixture

If $r$ is the number of independent reactions, the set of $(n-1)$ balance equations (8.1) can be transformed into a set of $r$ balance equations plus $m = n-1-r$ conservation equations. This transformation is particularly useful when $m$ is large.

Denote by $\underline{\underline{N}}$ the $(n \times r)$ matrix of the reduced stoichiometric coefficients

$$(8.12) \qquad \left(\underline{\underline{N}}\right)_{\ell i} = \nu_{\ell i} \quad \left[1 \leq \ell \leq r \; ; \; 1 \leq i \leq n\right]$$

each row pertaining to one reaction and each column pertaining to one component of the mixture.

By definition, since the number of independent reactions is $r$ , the rank of this matrix is equal to $r$ . Hence unless an inessential change in the order of columns, one can write $\underline{\underline{N}}$ in the partitioned form:

$$(8.13) \qquad \underline{\underline{N}} = \left[\underline{\underline{R}} \quad \underline{\underline{Q}}\right]$$

where $\underline{\underline{R}}$ is a non-singular square matrix of order $r$ and $\underline{\underline{Q}}$ is a matrix of order $(r \times m)$ (with $m = n-r$). Perform on the set

$(C_1, C_2, \ldots\ldots, C_n)$ the same permutation which was needed to obtain $\underline{\underline{R}}$ and denote by $\underline{C}_1, \underline{C}_2$ and $\underline{\xi}$ the column matrices defined by:

$$C_1 = \begin{bmatrix} C_1 \\ \vdots \\ C_r \end{bmatrix} \; ; \; C_2 = \begin{bmatrix} C_{r+1} \\ \vdots \\ C_n \end{bmatrix} \; ; \; \underline{\xi} = \begin{bmatrix} \xi_1 \\ \vdots \\ \xi_r \end{bmatrix} \tag{8.14}$$

The relation (2.11) between the virtual displacements $\delta C_i$ and $\delta \xi_\ell$ can also be written as:

$$\dot{C}_i = \sum_{\ell=1}^{r} \nu_{\ell i} \dot{\xi}_\ell \tag{8.15}$$

or, on account of eqs. (8.12), (8.13), (8.14) :

$$\begin{bmatrix} \dot{\underline{C}}_1 \\ \dot{\underline{C}}_2 \end{bmatrix} = \begin{bmatrix} \tilde{\underline{\underline{R}}} \\ \tilde{\underline{\underline{Q}}} \end{bmatrix} \cdot \dot{\underline{\xi}}$$

from which

$$\begin{aligned} \dot{\underline{C}}_1 &= \tilde{\underline{\underline{R}}} \cdot \dot{\underline{\xi}} \\ \dot{\underline{C}}_2 &= \tilde{\underline{\underline{Q}}} \cdot \dot{\underline{\xi}} = \tilde{\underline{\underline{Q}}} \cdot \left(\tilde{\underline{\underline{R}}}\right)^{-1} \cdot \dot{\underline{C}}_1 \end{aligned} \tag{8.16}$$

where a dot denotes matrix multiplication, a tilde the transpose of a matrix and $\underline{\underline{R}}^{-1}$ is the inverse of the non singular matrix $\underline{\underline{R}}$. Thus, as expectable, only r productions $\dot{\underline{C}}_i$ are independent.

Perform now the following linear transformation of the concentration $C_i$ :

$$\begin{bmatrix} \underline{\xi} \\ \\ \underline{\sigma} \end{bmatrix} = \underline{\underline{M}} \cdot \begin{bmatrix} (\underline{C}_1 - \underline{C}_{10}) \\ \\ (\underline{C}_2 - \underline{C}_{20}) \end{bmatrix} \tag{8.17}$$

where:

$$\underline{\sigma} = \begin{bmatrix} \sigma_1 \\ \\ \sigma_m \end{bmatrix} \qquad (m = n - r)$$

The column matrices $\underline{C}_{10}$ and $\underline{C}_{20}$ are constant matrices such that $\sum_{i=1}^{n} \underline{C}_{io} = 1$. The non singular square matrices $\underline{\underline{M}}$, of order $n$ is defined as:

$$\underline{\underline{M}} = \begin{bmatrix} \tilde{\underline{\underline{R}}}^{-1} & \underline{\underline{0}} \\ \\ -\tilde{\underline{\underline{Q}}} \cdot \tilde{\underline{\underline{R}}}^{-1} & \underline{\underline{\eta}} \end{bmatrix}$$

with $\underline{\underline{\eta}}$ the unit matrix of order $(m)$ and $\underline{\underline{0}}$ the null matrix of order $(r \times m)$. Since $\underline{\underline{M}}$ and $\underline{C}_{io}$ are constant matrices the productions $\dot{\underline{\xi}}$ and $\dot{\underline{\sigma}}$ of $\xi_i$ and $\sigma_i$ are given by:

$$\dot{\underline{\xi}} = \underline{\underline{\tilde{R}}}^{-1} \cdot \dot{\underline{C}}_1$$

$$\dot{\underline{\sigma}} = - \underline{\underline{\tilde{Q}}} \cdot \tilde{R}^{-1} \cdot \dot{\underline{C}}_1 + \dot{C}_2 = 0 \qquad (8.18)$$

where the last equality follows from the second of equation (8.16). Hence the quantities $\sigma_i \left[1 \leq i \leq m = n-r\right]$ are conservative quantities.

By performing the linear transformation (8.7) on equations $(8.11)_3$ one obtains:

$$\rho \frac{\partial \xi_\ell}{\partial t} + \underline{\nabla} \cdot \underline{j}_{\xi_\ell} = \dot{\xi}_\ell$$

$$\rho \frac{\partial \sigma_i}{\partial t} + \underline{\nabla} \cdot \underline{j}_{\sigma_i} = 0 \qquad (8.19)$$

where the new diffusive fluxes $\underline{j}_{\xi_\ell}$ and $\underline{j}_{\sigma_i}$ are combinations of the old ones $\underline{j}_{c_i}$ defined by:

$$\underline{j}_{\xi_\ell} = \sum_{s=1}^{r} \left(\underline{\underline{R}}^{-1}\right)_{s\ell} \underline{j}_{c_s} \qquad (1 \leq \ell \leq r)$$

$$\underline{j}_{\sigma_i} = \underline{j}_{c_{r+1}} - \sum_{s=1}^{r} \left(\underline{\underline{R}}^{-1} \cdot \underline{\underline{Q}}\right)_{si} \cdot \underline{j}_{c_s} ; \qquad (1 \leq i \leq m = n-r) \qquad (8.20)$$

When the parameters $\xi_\ell$ and $\eta_i$ are known, the mass concentrations are to be obtained from the inverse of the transformation (8.17). Upon a well known rule of matrical algebra it is:

(8.21) $$\underline{\underline{M}}^{-1} = \begin{bmatrix} \underline{\underline{R}} & \underline{\underline{0}} \\ \underline{\underline{Q}} & \underline{\underline{n}} \end{bmatrix}$$

so that [see eq. (8.15)] :

(8.22)
$$C_j - C_{jo} = \sum_{\ell=1}^{r} (\underline{\underline{R}})_{\ell j} = \sum_{\ell=1}^{r} \nu_{\ell j} \xi_\ell \; ; \quad 1 \leq j \leq r$$
$$C_{r+1} - C_{r+1,0} = \sigma_i + \sum_{\ell=1}^{r} (\underline{\underline{Q}})_{\ell j} \xi_\ell = \sigma_i + \sum_{\ell=1}^{r} \nu_{\ell i} \xi_\ell \; ; \; (1 \leq i \leq m = n-r)$$

Notice that the $m$-th "component" $\sigma_m$ of $\underline{\sigma}$ is not an independent quantity for:

$$0 = \sum_{j=1}^{r} (C_j - C_{jo}) + \sum_{i=1}^{m} (C_{r+i} - C_{r+i,o}) = \sum_{i=1}^{m} \sigma_i$$

since, upon the mass conservation:

$$\sum_{j=1}^{r} (\underline{\underline{R}})_{\ell j} + \sum_{i=1}^{m} (\underline{\underline{Q}})_{\ell i} = \sum_{i=1}^{n} \nu_{\ell i} = 0 \qquad (\forall \ell)$$

For a closed system, $\underline{j}_{ci} = 0$ $(\forall i)$ so that, upon the equations (8.19) $\underline{j}_{\xi\ell} \equiv 0$ ; $\underline{j}_{\sigma_i} \equiv 0$ $(\forall i,\ell)$ and the equations $(8.18)_2$ are readily integrated to give $\sigma_i \equiv 0$. The constants $C_{io}$ will then assume the meaning of the composition of the mixture for $\xi_\ell = 0$. The introduction of progress variables may have some advantages even for open systems.

The balance equations $(8.19)_2$ clearly express the conservation of the mass of all atomic species present in the mixture and the parameters $\sigma_i$ are proportional to the "total" mass concentration of these atomic species. They include the masses of atomic species both in their (eventual) free and combined forms.

When the chemical reactions are not all independent one first introduces the independent progress variables $\underline{\eta}$ as described in Example 3, and then applies the transformations formula (8.13).

Example 15

With reference to the system discussed in Example 3, the independent progress variables $\eta_i$ are defined by eqs. (2.10). Hence $n=6; r=4; m=2$ and upon eq. (2.9) as the matrix $\underline{\underline{N}}$ defined by eq. (8.12) we must take the matrix denoted by $\underline{\underline{N}}$, in Example 3, i.e. the matrix formed by the first four rows of the matrix given by eq. (2.5). We permute the order of its columns so as to bring the first two to the last two places. Hence the matrices $\underline{C}_1$ and $\underline{C}_2$ are given, with an obvious meaning of the symbols (see eq. (2.5), by:

$$(8.23) \qquad \underline{C}_1 = \begin{bmatrix} C_{H_2O} \\ C_{OH} \\ C_{O_2} \\ C_{H_2} \end{bmatrix} \qquad \underline{C}_2 = \begin{bmatrix} C_H \\ C_O \end{bmatrix}$$

and one has:

$$\tilde{\underline{\underline{R}}} = \begin{bmatrix} 0 & 0 & 18 & -18 \\ 17 & 17 & -34 & 17 \\ -32 & 0 & 0 & 0 \\ 0 & -2 & 0 & 0 \end{bmatrix}$$

$$\tilde{\underline{\underline{Q}}} = \begin{bmatrix} 1 & 1 & 0 & 1 \\ 16 & 16 & 16 & 0 \end{bmatrix}$$

from which:

$$\tilde{\underline{\underline{R}}}^{-1} = \begin{bmatrix} 0 & 0 & -\frac{1}{32} & 0 \\ 0 & 0 & 0 & -\frac{1}{2} \\ -\frac{1}{18} & -\frac{1}{17} & -\frac{1}{32} & -\frac{1}{2} \\ -\frac{1}{9} & -\frac{1}{17} & -\frac{1}{32} & -\frac{1}{2} \end{bmatrix}, \quad \tilde{\underline{\underline{Q}}} \cdot \tilde{\underline{\underline{R}}}^{-1} = \begin{bmatrix} \frac{1}{9} & \frac{1}{17} & 0 & 1 \\ \frac{16}{18} & \frac{16}{17} & 1 & 0 \end{bmatrix}$$

Thus, according to equations (8.17) and letting, for simplicity, $C_{io} = 0 \ (\forall i)$, the new variable $\underline{\sigma} \equiv (\sigma_1, \sigma_2)$ is defined by:

$$\begin{bmatrix} \sigma_1 \\ \sigma_2 \end{bmatrix} = \begin{bmatrix} C_H \\ C_O \end{bmatrix} + \begin{bmatrix} \frac{1}{9} & \frac{1}{17} & 0 & 1 \\ \frac{16}{18} & \frac{16}{17} & 1 & 0 \end{bmatrix} \cdot \begin{bmatrix} C_{H_2O} \\ C_{OH} \\ C_{O_2} \\ C_{H_2} \end{bmatrix}$$

or

$$\sigma_1 = C_H + \frac{1}{9} C_{H_2O} + \frac{1}{17} C_{OH} + C_{H_2}$$

$$\sigma_2 = C_O + \frac{16}{18} C_{H_2O} + \frac{16}{17} C_{OH} + C_{O_2}$$

It appears that $\sigma_1$ , and $\sigma_2$ are the total mass concentration of $H$ and $O$ , respectively, since for instance, the term $\frac{1}{9} C_{H_2O}$ represents the contribution of $H$ ,due to the presence of the "compound" $H_2O$. Indeed, let $m_H^{(H_2O)}$ be the mass of $H$ present in the combined form $(H_2O)$and $m_{H_2O}$ the mass of $H_2O$. For the moles $n_{H_2O}$ , $m_{H_2O} = n_{H_2O}(2m_H + m_O)$, the mass of $H$ is $2m_H n_{H_2O}$ if, as usual , $m_i$ denotes the molar of the $i$-th element Then:

$$\frac{m_H^{(H_2O)}}{m_{H_2O}} = \frac{2m_H}{2m_H + m_O} = \frac{1}{9}$$

since $m_O / m_H = 16$. Hence (1.3) $C_{H_2O}$ is the concentration of $H$ in its "combined" form $H_2O$ .

## 9. MOMENTUM EQUATIONS

We must now formulate balance equations for the linear and angular momenta. These are the only independent balance equations of kinematic quantities in the subject single-fluid theory. [In this theory one assumes that the kinematic state of the medium is adequately characterized by a single mass velocity $\underline{V}$ . When dealing with mixtures (in particular multiphase systems) it may prove necessary to consider as basic unknowns the mass velocities $\underline{V}_i$ of each constituent. In this case one would have to formulate independent momentum balance equations for each component]. As before, we shall deal explicitly only with the differential and jump forms of the balance equation.

For the linear momentum we can write, formally:

$$(9.1) \qquad \frac{\partial \varrho \underline{V}}{\partial t} + \underline{\nabla} \cdot \underline{\underline{\pi}} = \underline{\dot{f}} = \varrho \underline{g}$$

where: $\varrho \underline{V}$ , by the very same definition of $\underline{V}$ , is the momentum density in the Galileian frame $G$, $\pi$ is the total (tensorial) momentum flux and $\underline{\dot{f}}$ is the production of linear momentum per unit volume. In the assumed absence of charged particles, this force is simply the one due to the earth gravitational field. Hence $\underline{\dot{f}} = \varrho \underline{g}$ where $\underline{g}$ is the gravitational acceleration vector. In many cases the contribution due to the gravitational field is negligible compared to that due to the convection of momentum.

As done with the vectorial fluxes, we can express the total flux of momentum $\underline{\underline{\pi}}$ as the sum of a convective flux $\varrho \underline{V}\,\underline{V}$ plus a diffusive flux $\underline{\underline{r}}$

$$\underline{\underline{\pi}} = \varrho \underline{V}\,\underline{V} + \underline{\underline{r}} \tag{9.2}$$

The tensor $\underline{\underline{r}}$ is Galileian invariant. Indeed, from the Galileian invariance of the momentum equation and of $\underline{\underline{q}}$ it follows that:

$$\frac{\partial \varrho \underline{V}}{\partial t} + \underline{\nabla} \cdot \underline{\underline{\pi}} = \frac{\partial \varrho \underline{V}'}{\partial t'} + \underline{\nabla} \cdot \underline{\underline{\pi}}'$$

On the other hand, when accounting for the continuity equation

$$\frac{\partial \varrho \underline{V}'}{\partial t'} = \frac{\partial \varrho \underline{V}'}{\partial t} + \underline{C} \cdot \nabla (\varrho \underline{V}') = \frac{\partial \varrho (\underline{V} - \underline{C})}{\partial t} + \underline{C} \cdot \nabla \left[ \varrho (\underline{V} - \underline{C}) \right] =$$

$$= \frac{\partial \varrho \underline{V}}{\partial t} + \underline{C}\, \underline{\nabla} \cdot (\varrho \underline{V}) + \underline{C} \cdot \nabla (\varrho \underline{V}) - \underline{\nabla} \cdot (\varrho \underline{C}\,\underline{C})$$

from which:

$$\frac{\partial \varrho \underline{V}}{\partial t'} = \frac{\partial \varrho \underline{V}}{\partial t} + \underline{\nabla} \cdot \left[ \varrho \left( \underline{V}\, \underline{C} + \underline{C}\, \underline{V} - \underline{C}\,\underline{C} \right) \right]$$

since (recall that $\underline{C}$ is a constant velocity)

$$\underline{C}\, \underline{\nabla} \cdot (\varrho \underline{V}) = \underline{\nabla} \cdot \left[ \varrho \underline{V}\, \underline{C} \right]$$

Hence:

$$\frac{\partial \varrho V'}{\partial t'} - \frac{\partial \varrho \underline{V}}{\partial t} = \underline{\nabla} \cdot \left[ \underline{\underline{\pi}} - \underline{\underline{\pi}}' \right] =$$

$$= \underline{\nabla} \cdot \left[ \varrho \left( \underline{V}\,\underline{C} + \underline{C}\,\underline{V} - \underline{C}\,\underline{C} \right) \right] = \underline{\nabla} \cdot \left[ \varrho \left( \underline{V}\,\underline{V} - \underline{V}'\underline{V}' \right) \right]$$

Thus:

$$\underline{\underline{\pi}} - \varrho \underline{V}\,\underline{V} = \underline{\underline{\tau}} = \underline{\underline{\pi}}' - \varrho \underline{V}'\underline{V}' = \tau \quad \text{(q.e.d)}$$

If the fluid is a mixture $\varrho\underline{V}\,\underline{V}$ is <u>not</u> the total convective momentum flux. Consider, indeed, for simplicity, a binary mixture and let $\varrho_1 \underline{V}_1$ and $\varrho_2 \underline{V}_2$ be the momentum densities of the two components of the mixture. The total momentum density is given by $\varrho\underline{V} = \varrho_1\underline{V}_1 + \varrho_2 \underline{V}_2$ (with $\varrho = \varrho_1 + \varrho_2$ if both components are gas). The mass velocity $\underline{V}$ is the velocity of the center of gravity of the mixture while the $\underline{V}_i$ are those of the two constituents. The convection of total momentum is given by $(\varrho_1 \underline{V}_1 \underline{V}_1 + \varrho_2 \underline{V}_2 \underline{V}_2)$ and it is:

$$\varrho_1 \underline{V}_1 \underline{V}_1 + \varrho_2 \underline{V}_2 \underline{V}_2 = \varrho \underline{V}\,\underline{V} + \frac{\varrho_1 \varrho_2}{\varrho} \underline{W}\,\underline{W}$$

where $\underline{W} = \underline{V}_2 - \underline{V}_1$ is the relative mass velocity between the two constituent and, clearly:

$$\underline{V}_1 = \underline{V} - \frac{\varrho_2}{\varrho} \underline{W}$$

$$\underline{V}_2 = \underline{V} - \frac{\varrho_1}{\varrho} \underline{W}$$

Thus, as said, $\varrho \underline{V}\,\underline{V}$ is not the total convective momentum flux. The latter contains, in addition, the contribution $\left(\frac{\varrho_1 \varrho_2}{\varrho} \underline{W}\,\underline{W}\right)$ giving the convective flux of total momentum measured in the frame moving with the velocity $\underline{V}$ (this contribution is, obviously, Galileian invariant).

The term $\frac{\varrho_1 \varrho_2}{\varrho} \underline{W}\,\underline{W}$, being of the order of the square of the relative velocity $\underline{W} = \underline{V}_2 - \underline{V}_1$, is almost always negligible. [When this is no longer so one must revert to a multi-fluid theory]. To within this approximation $\underline{\underline{\tau}}$ can be interpreted as the stress tensor of the fluid.

It proves convenient, for the further developments, to seperate from $\underline{\underline{\tau}}$ an isentropic part by letting:

$$\underline{\underline{\tau}} = \pi\, \underline{\underline{n}} + \underline{\underline{\tau}}_0 \tag{9.3}$$

where $\underline{\underline{n}}$ is the unit tensor, $\underline{\underline{\tau}}_0$ the divergenceless part of $\underline{\underline{\tau}}$ (i.e. its deviatoric part) and:

$$\pi = \frac{1}{3} \mathrm{Tr}\left(\underline{\underline{\tau}}\right) \tag{9.4}$$

The scalar $\pi$ is thus equal to one third of the linear invariant of $\tau$ .(i.e. $\pi$ is the average normal stress). Obviously $\pi$ is Galileian invariant.

When one lets:

$$\underline{\underline{\pi}} = \varrho \underline{V}\,\underline{V} + \pi\, \underline{\underline{n}} + \underline{\underline{\tau}}_0 \tag{9.5}$$

the momentum equation can also be written in terms of the operator $\frac{D}{Dt}$ as:

$$\varrho \frac{D\underline{V}}{Dt} + \underline{\nabla} \cdot \left[\pi \underline{\underline{n}} + \tau_0\right] = \varrho \underline{g} \tag{9.6}$$

As with the scalar balance equations, each term of this equation is, individually, Galileian invariant.

With the sign convention used in equation (9.6) the normal stresses (and hence $\pi$ ) are positive if they are compressive stresses.

We should now formulate the balance equation for angular momentum. As known, the total angular momentum of a system with respect to any point O can be expressed as sum of the moment with respect to O , of the motion of the c.g. of the system (in this case the elementary particle) plus an "intrinsic" momentum, associated with the motion around the c.g. of the system.

The balance equation for the first part of the angular momentum is readily obtained from the equation for the linear momentum by multiplying it vectorially by the radius vector r stemming from the given point O .

Indeed, since $\frac{D\underline{r}}{Dt} = \underline{V}$, by definition, one gets that:

$$\varrho \frac{D}{Dt}\left(\underline{r} \wedge \underline{V}\right) = \varrho \underline{r} \wedge \frac{D\underline{V}}{Dt}$$

Hence:

$$\varrho \frac{D}{Dt}\left[\underline{r} \wedge \underline{V}\right] + \underline{r} \wedge \left[\underline{\nabla} \cdot \left(\pi \underline{\underline{n}} + \underline{\underline{\tau}}_0\right)\right] = \varrho \underline{r} \wedge \underline{g} \tag{9.7}$$

This balance equation does not represent an independent statement since it will be identifically satisfied whenever the balance equation for linear momentum is. Notice that the quantity

$$-\underline{\nabla} \cdot \left[\pi \underline{\underline{n}} + \underline{\underline{\tau}}_0\right]$$

represents the force acting, per unit volume, on the elementary particle as resulting from the surface stresses distributed over its bounding elementary volume. Thus eq. (9.7) simply expresses the fact that the time rate of change of the momentum of the motion of the c.g. equals the moment of the total forces acting on the elementary particle, considered as a rigid body.

It can be shown that, whenever one can make the assumption of negligible intrinsic couple stress density (which is quite acceptable for normal fluids), the balance of intrinsic angular momentum leads to the condition that the stress tensor $\underline{\underline{\tau}}_0$ must be symmetric. Hence, by imposing the symmetry of $\underline{\underline{\tau}}_0$ the balance of angular momentum is identically satisfied (if that of linear momentum is) and need not be formulated as an independent statement. Equivalently, on denoting by $\tilde{\underline{\underline{\tau}}}_0$ the transpose of $\underline{\underline{\tau}}$ , one may say that the balance of angular momentum reduces to the equation:

expressing the symmetry of the stress tensor.

## 10. ENERGY CONSERVATION EQUATION. BALANCE EQUATIONS FOR INTERNAL, KINETIC AND POTENTIAL ENERGIES. HEAT FLUXES.

The total energy per unit mass (e) in G is given by the sum of the specific internal energy ($u$) of the mixture plus its potential energy $\psi$ (per unit mass) plus its kinetic energy (per unit mass). The latter, in the present one-fluid theory, coincides with the kinetic energy ($V^2/2$) connected with the mass motion of the mixture [i.e. in the one-fluid theory one neglects the kinetic energies per unit mass of the mixture, $C_i V_i^2/2$, corresponding to the mass motions of the components of the mixture [Ref. 3, 8] .

This total energy is conserved upon a basic postulate. Thus if $\underline{j}_e$ denotes the total diffusive flux (Galileian invariant) one writes:

$$\rho \frac{D}{Dt}\left(u + \frac{V^2}{2} + \psi\right) + \underline{\nabla} \cdot \underline{j}_e = 0 \tag{10.1}$$

Balance equations for each of the three forms of energy can also be formulated. Upon the <u>shifting equilibrium hypothesis</u>, the balance equation for the internal energy can be expressed in terms of those of its natural variables. The balance equation for kinetic energy is obtained from the scalar multiplication of the momentum equation by $\underline{V}$ . The balance equation for $\psi$ follows from its very definition. When combining these relations with equations (10.1) one obtains the expressions for $\underline{j}_e$ and for the entropy production.

10.1 Balance of potential energy

By definition of the earth gravitational potential $\psi$ it is:

$$\underline{\nabla}\,\psi = -\underline{g}$$

If we make the more than plausible assumption that the potential does not depend explicitly on the time we get:

$$\varrho\,\frac{D\psi}{Dt} = -\varrho\,\underline{V}\cdot\underline{g} \tag{10.2}$$

which represents the required balance equation for the potential energy. Quite obviously, the production of potential energy equals to minus the power density (i.e. power per unit volume) associated with the work done by the body force.

10.2 Balance of internal energy

For the balance of internal energy one gets, upon the Gibbs relation $(1.6)_2$ and the balance equations (8.8):

$$\varrho\,\frac{Du}{Dt} = \dot{u} - \underline{\nabla}\cdot\underline{j}_u = \varrho\left[T\,\frac{Ds}{Dt} - p\,\frac{Dv}{Dt} + \right.$$

$$\left. + \sum_{i=1}^{n}(g_i - g_n)\,\frac{DC_i}{Dt} + \sum_{j=1}^{K}(\Theta_j - T)\,\frac{Ds_j}{Dt}\right] = \tag{10.3}$$

$$= T\left(\dot{s} - \underline{\nabla}\cdot\underline{j}_s\right) - p\,\underline{\nabla}\cdot\underline{V} + \sum_{i=1}^{n-1}\left(g_i - g_n\right)\left(\dot{c}_i - \underline{\nabla}\cdot\underline{j}_{c_i}\right) +$$

$$+ \sum_{j=1}^{k}\left(\Theta_j - T\right)\left(\dot{s}_j - \underline{\nabla}\cdot\underline{j}_{s_j}\right) \tag{10.3}$$

We shall dispose of the often mentioned arbitrariness in the definition of fluxes and productions by requiring that the diffusive flux be a linear operator (*). From the Euler relation $(1.4)_2$ it then follows that (recall that $\underline{j}_v = 0$)

$$\underline{j}_u = T\underline{j}_s + \sum_{i=1}^{n} g_i\underline{j}_{c_i} + \sum_{j=1}\left(\Theta_j - T\right)\underline{j}_{s_j} \tag{10.4}$$

and comparison with eq. (10.3) yields:

$$\dot{u} = T\dot{s} - p\,\underline{\nabla}\cdot\underline{V} + \sum_{i=1}^{n-1}\left(g_i - g_n\right)\dot{c}_i + \sum_{j=1}^{k}\left(\Theta_j - T\right)\dot{s}_j + \underline{j}_s\cdot\underline{\nabla}T +$$

$$+ \sum_{i=1}^{n-1}\underline{j}_{c_i}\cdot\underline{\nabla}\left(g_i - g_n\right) + \sum_{j=1}^{k}\underline{j}_{s_j}\cdot\nabla\left(\Theta_j - T\right) \tag{10.5}$$

(*) It can be shown (Ref.10) that this is the only assumption consistent with the first order homogeneity of the internal energy fundamental relation.

where, on account of eqs. (8.15), the third term can be replaced by :

$$\sum_{l=1}^{r} A_l \dot{\xi}_l$$

if $r$ is the number of independent chemical reactions.

10.3 Balance equation for the kinetic energy

As said, this balance equation is obtained from the momentum equation:

$$\rho \frac{D\underline{V}}{Dt} + \underline{\nabla} \cdot \left[ \pi \underline{\underline{u}} + \underline{\underline{\tau}}_0 \right] = \rho \underline{g}$$

multiplied scalarly by $\underline{V}$ since:

$$\rho \frac{D\, V^2/2}{Dt} = \rho \underline{V} \cdot \frac{D\underline{V}}{Dt}$$

On account of the symmetry of $\underline{\underline{\tau}} = \underline{\underline{\pi}} + \underline{\underline{\tau}}_0$ it is:

$$\underline{V} \cdot \left[ \underline{\nabla} \cdot \underline{\underline{\tau}} \right] = \underline{\nabla} \cdot \left[ \underline{\underline{\tau}} \cdot \underline{V} \right] - \underline{\underline{\tau}} : \underline{\nabla}\, \underline{V}$$

The dyadic ( $\underline{\nabla}\,\underline{V}$ ) can be expressed as sum of a symmetric part $(\underline{\nabla}\,\underline{V})_s$ and an anti-symmetric one $(\underline{\nabla}\,\underline{V})_a$ . The double scalar multiplication of the latter one by the symmetric tensor $\underline{\underline{\tau}}$ is equal to zero, so that:

$$\underline{\underline{\tau}} : \underline{\nabla}\,\underline{V} = \underline{\underline{\tau}} : (\underline{\nabla}\,\underline{V})_s$$

Physically, the antisymmetric tensor $(\underline{\nabla}\,\underline{V})_a$ is equivalent to an axial vector which is proportional to the average angular velocity of the particle. Since a symmetric distribution of stresses perform no work for an angular rotation of the particle the quantity $\underline{\underline{\tau}} : (\underline{\nabla}\,\underline{V})_a = 0$.

Furthermore, since $\underline{\underline{u}} : \underline{\nabla}\,\underline{V} \equiv \underline{\nabla} \cdot \underline{V}$ it is

$$\underline{\underline{\tau}} : (\underline{\nabla}\,\underline{V})_s = \left[\pi\,\underline{\underline{u}} + \underline{\underline{\tau}}_0\right] : (\underline{\nabla}\,\underline{V})_s =$$

$$= \pi\,(\underline{\nabla} \cdot \underline{V}) + \underline{\underline{\tau}}_0 : (\underline{\nabla}\,\underline{V})_0$$

where $(\underline{\nabla}\,\underline{V})_0$ is the symmetric, traceless part of ( $\underline{\underline{\tau}}$ ).

i.e.

$$(\underline{\nabla}\,\underline{V})_0 = (\underline{\nabla}\,\underline{V})_s - \frac{1}{3}(\underline{\nabla} \cdot \underline{V})\,\underline{\underline{u}}$$

Hence, finally, the balance of kinetic energy reads:

$$\varrho\,\frac{D V^2/2}{Dt} + \underline{\nabla} \cdot \left[\underline{\underline{\tau}} \cdot V\right] = \underline{\underline{\tau}} : \underline{\nabla}\,\underline{V} + \varrho\,\underline{g} \cdot \underline{V} =$$

$$= \pi\,\underline{\nabla} \cdot \underline{V} + \underline{\underline{\tau}}_0 : (\underline{\nabla}\,\underline{V})_0 + \varrho\,\underline{g} \cdot \underline{V} \qquad (10.6)$$

Thus:

- The (diffusive) flux of kinetic energy is equal to the work done, per unit time, by the stress tensor $\underline{\underline{\tau}}$
- The production of kinetic energy is made up of two parts, one

is associated with the "rate of deformation" of the elementary particle and the other is equal to minus the production of potential energy.

It follows that:

a) potential energy is converted directly and entirely into kinetic energy, in the sense that to an increase (resp. decrease) of potential energy corresponds an equal decrease (resp. increase) of kinetic energy [i.e. the contribution $\varrho \underline{g} \cdot \underline{V}$ in the production of kinetic energy can have any sign].

b) the converse is not true. Specifically, kinetic energy can be converted into potential and/or internal energy. Indeed, since the production of total energy must vanish one obtains that

$$\dot{u} = -\underline{\underline{\tau}} : \underline{\nabla}\,\underline{V} = -\pi(\underline{\nabla} \cdot \underline{V}) - \underline{\underline{\tau}}_0 : (\underline{\nabla}\,\underline{V})_0 \qquad (10.7)$$

We shall come back to this expression after having obtained the expression for the entropy production and imposed the postulate of positive entropy production.

Since the diffusive flux of potential energy is zero, the diffusive flux of total energy is simply the sum the of those of internal and kinetic energies:

$$\underline{j}_e = \underline{j}_u + \underline{\underline{\tau}} \cdot \underline{V} = T\,\underline{j}_s + \sum_{i=1}^{n-1} (g_i - g_n)\,\underline{j}_{c_i} + \qquad (10.8)$$

$$(10.8) \qquad + \sum_{i=1}^{k} \left(\Theta_i - T\right) \underline{j}_{s_i} + \underline{\underline{\tau}} \cdot V$$

## 10.4 Definitions of heat fluxes

For closed systems in thermal equilibrium the definition of heat-flux is unique. This is not so with open systems and, a fortiori, with open systems which are not in thermal equilibrium.

This arbitrariness in the definition of heat flux stems from the fact that any time there is a flux of matter or of entropy there is also a flux of energy [as equation (10.8) clearly shows] and it is only a matter of convention (or of convenience) to single out one or more of these fluxes of energy and call them "heat-flux".

There are thus several choices possible for the definition of heat-flux. To any particular choice will respond, as we shall see, a special form for the expression of the entropy production and for the energy conservation equation. Any possible defintion obviously leaves all physical results unchanged and one should only pay attention in using it consistently, i.e. in associating to it the pertinent expression for the entropy production and energy conservation equation.

We shall discuss only two such definitions.

The first one is, so to speak, a direct one in so far as one simply says that the flux of total energy $\underline{j}_e$ is made up of two contributions: a work-flux [which is clearly given by the term $\underline{\underline{\tau}} \cdot \underline{V}$ in equation (10.8)] and a heat-flux. In other words one <u>defines</u> as heat flux $\underline{j}_q$ the flux of internal energy. Thus:

$$\underline{j}_q \equiv \underline{j}_u = T\underline{j}_s + \sum_{i=1}^{n-1} (g_i - g_n)\underline{j}_{c_i} + \sum_{j=1}^{k} (\Theta_j - T)\underline{j}_{s_j} \qquad (10.9)$$

As said, any time there is a diffusion of mass there is also an associated diffusion of energy and entropy. In the second definition, the first contribution is "eliminated" from the heat-flux. flux. The diffusive energy flux associated with mass diffusion can be written, on account of the relation

$$\sum_{i=1}^{n} \underline{j}_{c_i} = 0$$

and of eq. (10.8), as:

$$\sum_{i=1}^{n} g_i \underline{j}_{c_i}$$

This flux is split in two contributions with the help of the thermodynamic equality (5.11):

$$\underset{\sim}{\hat{H}}_i = T\hat{S}_i + g_i$$

where $\hat{S}_i$ and $\underset{\sim}{\hat{H}}_i$ are, respectively, the partial specific entropy and specific generalized enthalpy (recall that the latter one is equal to the partial specific enthalpy in thermal equilibrium) of component $i$ . Hence:

$$\sum_{i=1}^{n} g_i \underline{j}_{c_i} = \sum_{i=1}^{n} \left[ \hat{\underset{\sim}{H}}_i \underline{j}_{c_i} - T \hat{S}_i \underline{j}_{c_i} \right]$$

and the second definition of heat flux $\underline{j}'_q$ is:

$$(10.10) \qquad \underline{j}'_q = \underline{j}_q - \sum_{i=1}^{n} \hat{\underset{\sim}{H}}_i \underline{j}_{c_i} = \underline{j}_u - \sum_{i=1}^{n} \hat{\underset{\sim}{H}}_i \underline{j}_{c_i} .$$

This definition coincides with the one usually given when the mixture is in thermal equilibrium. With either definition of the heat flux, the flux of entropy is no longer an independent flux. In terms of $\underline{j}_q$ and $\underline{j}'_q$ one has, from eqs. (10.8),(10.9) and (10.10):

$$(10.11) \qquad \begin{aligned} \underline{j}_s &= \frac{1}{T} \left[ \underline{j}_q - \sum_{i=1}^{n} g_i \underline{j}_{c_i} - \sum_{j=1}^{n} (\Theta_i - T) \underline{j}_{s_j} \right] \\ \underline{j}_s &= \frac{1}{T} \underline{j}'_q + \sum_{i=1}^{n} \hat{S}_i \underline{j}_{c_i} - \sum_{j=1}^{k} \frac{(\Theta_i - T)}{T} \underline{j}_{s_j} \end{aligned}$$

As said, the term $\underline{\underline{\tau}} \cdot \underline{V}$ is the work-flux contribution to the total diffusive flux of energy $\underline{j}_e$ . This contribution can also be written as

$$(10.12) \qquad \underline{\underline{\tau}} \cdot \underline{V} = p \underline{V} + \left[ (\pi - p) \underline{V} + \underline{\underline{\tau}}_0 \cdot \underline{V} \right]$$

where $p$ is the thermodynamic pressure. As we shall see, in the next paragraph, only the terms in the squared brackets contribute to the production of entropy.

Thus they represent the irreversible part of the work-flux. The reversible part ($p\underline{V}$) can be combined with the convective flux of total energy. On account of the definition of the thermodynamic enthalpy:

$$h = u + pv$$

of the expression (10.8) for $\underline{j}_e$ and of eq. (10.12) one can indeed write the energy conservation equation as:

$$\varrho \frac{D}{Dt}\left(h + \frac{V^2}{2}\right) - \frac{\partial p}{\partial t} + \underline{\nabla}\cdot\left[\underline{j}_u + (\pi - p)\underline{V} + \underline{\underline{\tau}}_0\cdot V\right] = 0$$

a form of which is particularly useful in steady motion.

For later use we shall also write the two different forms of this energy conservation equation corresponding to the different definitions of heat-flux.

With obvious substitutions one gets:

$$\varrho \frac{D}{Dt}\left[h + \frac{V^2}{2}\right] - \frac{\partial p}{\partial t} + \underline{\nabla}\cdot\left[\underline{j}_q + (\pi - p)\underline{V} + \underline{\underline{\tau}}_0\cdot\underline{V}\right] = 0$$

(10.12)

$$\varrho \frac{D}{Dt}\left[h + \frac{V^2}{2}\right] - \frac{\partial p}{\partial t} + \underline{\nabla}\cdot\left[\underline{j}'_q + \sum_{i=1}^{n}\hat{\underset{\sim}{H}}_i\,\underline{j}_{c_i} + (\pi - p)\underline{V} + \underline{\underline{\tau}}_0\cdot\underline{V}\right] = 0$$

Notice, explicitly, that in conditions of thermal non-equilibrium the partial generalized enthalpy of the $i$-th constituent appears in this equation and not the usual partial specific enthalpy.

## 11. ENTROPY PRODUCTION

It is appropriate to summarize what we have done so far. Subject to the "shifting equilibrium" hypothesis and to within the limits of a "single fluid" theory we have assumed that the set of "primitive" unknowns is given by the $d+1=n+k+2$ specific quantities (one of which vectorial):

$$s\,;\ v\,;\ c_i\,,\left(1\leq i\leq n-1\right);\ s_j\,,\left(1\leq j\leq k\right);\ \underline{V} \tag{11.1}$$

where $n$ is the number of components of the gas mixture, $k$ the number of subsystems which are not in thermal equilibrium and $\underline{V}$ the total momentum per unit mass of the mixture.

We have then formulated a set of balance equations for the primitive unknowns. Each balance equation for an extensive quantity introduces two new unknowns : the production and the flux of the quantity. Since the total mass flux $J_m=\rho\underline{V}$ is also the total momentum density, i.e. one of the 'primitive' unknowns (*)

(*) More generally, in a multi-fluid theory, one assumes each mass flux as a "primitive" unknown. By formulating for each of them a balance equation one introduces, correspondingly, a stress tensors and a "body" force density (momentum production) for each fluid.

the set of $(d+1)$ balance equations introduces a total of $(2d+1)$ additional unknowns, which we have called "induced" unknowns.

Specifically, those "induced" unknowns are the $(d-1)=n+k$ vectorial diffusive fluxes of thermodynamic state variables:

$$\underline{j}_s \ ; \ \underline{j}_{c_i} , \ (1 \leq i \leq n-1); \ \underline{j}_{s_j} , \ (1 \leq j \leq k)$$

the $(d=n+k+1)$ productions of thermodynamic state variables:

$$\dot{s} \ ; \ \dot{v} \ ; \ \dot{c}_i , \ (1 \leq i \leq n-1) \ ; \ \dot{s}_j , \ (1 \leq j \leq k)$$

the tensorial flux $\underline{\underline{\tau}}$ and vectorial production $(\dot{\underline{f}})$ of momentum. Since $\underline{\underline{\tau}}$ amounts, in general, to nine scalar unknowns, the total number of scalar unknowns "induced" by the balance equations is $3(d-1)+d+9+3=4d+9$ . The description of the gravitational field and the imposition of conservation principles reduce the number of "induced" scalar unknowns to $(4d+1)$. In particular, the imposition of mass conservation relates $\dot{v}$ to to the divergence of the "primitive" unknown $\underline{V}$ ,that of intrinsic moment of momentum implies that $\underline{\underline{\tau}}$ is symmetric, the description of the gravitational field yields $\dot{\underline{f}}=\varrho\underline{g}$ with $\underline{g}$ known, and finally, the imposition of total energy conservation <u>plus</u> the shifting equilibrium hypothesis lead to the expression for the entropy productions $\dot{s}$ in terms of the remaining "induced" unknowns. This expression involves also the thermodynamic intensive parameters which, however, are to be considered as known functions of the primitive set of thermodynamic variables via a fundamental equation or, equivalently, $(d)$ independent equations of state. By separating from $\underline{\underline{\tau}}$ its isotropic part $(\pi \underline{\underline{u}})$ and referring to it the thermodynamic pressure $(p)$, the induced remaining unknowns may be listed as:

a) one tensorial, symmetric, traceless tensor:

(11.2a) $$\underline{\underline{\tau}}_o$$

b) $(d-1)=(n+k)$ vectorial fluxes:

(11.2b) $$\underline{j}_s\,;\ \underline{j}_{c_i}\ (1 \le i \le n-1)\,;\ \underline{j}_{s_j}\ (1 \le j \le k)$$

[or any combination thereof, such as those obtained by replacing the diffusive heat fluxes $\underline{j}_q$ (or $\underline{j}'_q$) or the diffusive mass fluxes $\underline{j}_{c_i}$ with the fluxes $\underline{j}_{\xi_e}, \underline{j}_{\sigma_i}$ defined in eqs. (8.10)]

c) $(d-1)=(n+k)$ scalar productions:

(11.2c) $$(\pi - p)\,;\ \dot{c}_i\ (1 \le i \le n-1)\,;\ \dot{s}_j\ (1 \le j \le k)$$

where, however, only $r$ productions $\dot{c}_i$ are independent if $r$ is the number of independent reactions. When we want to stress this fact we shall use in place of the $(n-1)$ linearly dependent productions $\dot{c}_i$ the $r$ independent productions $\dot{\xi}_\ell$ where $\xi_\ell$ is the progress variable. [In the case of a simple gas the induced unknowns reduce to $\underline{\underline{\tau}}_o, (\pi - p), \underline{j}_q = T\underline{j}_s = \underline{j}_u$].

To unify the terminology we shall refer to <u>all</u>

"induced" unknowns (11.2) as "generalized fluxes" and distinguish them according to their tensorial order. Thus the productions (11.2c) shall be referred to, from now on, as "scalar generalized fluxes".

Several different (albeit equivalent) forms can be given to the expression for the entropy production.

A first form is obtained by eliminating the production of internal energy $\dot{u}$ between equations (10.5) and (10.7) and reads:

$$-T\dot{s} = (\pi - p)\,\underline{\nabla}\cdot\underline{V} + \sum_{i=1}^{n-1}(g_i - g_n)\dot{c}_i + \sum_{j=1}^{k}(\Theta_j - T)\dot{s}_j +$$

$$+ \underline{\underline{\tau}}_0 : (\underline{\nabla}\,\underline{V})_0 + \left\{\underline{j}_s\cdot\underline{\nabla}T + \sum_{i=1}^{n}\underline{j}_{c_i}\cdot\underline{\nabla}g_i + \sum_{j=1}^{k}\underline{j}_{s_j}\cdot\underline{\nabla}(\Theta_j - T)\right\} =$$

$$= -T\left(\dot{s}_0 + \dot{s}_1 + \dot{s}_2\right) \qquad (11.3)$$

where, account has been taken of the fact that $\sum_{i=1}^{n}\underline{j}_{c_i} = 0$ and it has been denoted by $\dot{s}_i$ the production involving the generalized fluxes of order $(i)$.

Two other forms correspond to changes in the expression of $\dot{s}_1$

$$-T\dot{s}_1 = \underline{j}_s\cdot\underline{\nabla}T + \sum_{j=1}^{n}\underline{j}_{c_i}\cdot\underline{\nabla}g + \sum_{j=1}^{k}\underline{j}_{s_j}\cdot\underline{\nabla}(\Theta_j - T) \qquad (11.4)$$

which result from the replacement of the entropy diffusive flux $\underline{j}_s$ with the heat fluxes $\underline{j}_q$ and $\underline{j}'_q$ respectively.

When $\underline{j}_q$ is used, substituting eq. $(10.11)_1$ into eq. (11.4) yields:

$$-T\dot{s}_1 = \underline{j}_q \cdot \frac{\underline{\nabla} T}{T} + \sum_{i=1}^{n} \underline{j}_{c_i} \cdot \nabla\left(\frac{g_i}{T}\right) + \sum_{j=1}^{k} \underline{j}_{s_j} \cdot T \underline{\nabla}\left(\frac{\Theta_j - T}{T}\right). \tag{11.5}$$

When $\underline{j}'_q$ is used one must substitute for $\underline{j}_s$ the expression given by equation $(10.11)_2$ . The diffusive mass fluxes $\underline{j}_{c_i}$ will then be multiplied by the expression ( $\underline{\nabla} g_i + \dot{s}_i \underline{\nabla} T$ ) which, on account of the thermodynamic equalities (5.11) and (5.12), becomes successively:

$$\underline{\nabla} g_i + \hat{s}_i \,\underline{\nabla} T = \underline{\nabla} g_i + \frac{\underline{\nabla} T}{T}\left[\hat{h}_i - g_i\right] =$$

$$= T\,\underline{\nabla}\left(\frac{g_i}{T}\right) + \frac{h_i}{T}\,\underline{\nabla} T = \underline{\nabla}_T g_i \tag{11.6}$$

where $\underline{\nabla}_T g_i$ denotes the isothermic gradient of $g_i$ ,i.e. the gradient for T=const. The expression for the entropy production $\dot{s}_1$ becomes

$$-T\dot{s}_1 = \underline{j}'_q \cdot \frac{\underline{\nabla} T}{T} + \sum_{i=1}^{n} \underline{j}_{c_i} \cdot \underline{\nabla}_T (g_i) + \sum_{j=1}^{k} \underline{j}_{s_j} \cdot T \underline{\nabla}\left(\frac{\Theta_j - T}{T}\right). \tag{11.7}$$

Upon a basic postulate, entropy can only be produced. Thus:

$$\dot{s} \geq 0$$

The validity of any of the above expressions for the entropy production rests on the validity of the shifting-equilibrium hypothesis.

The quantities which multiply the generalized fluxes in eqs. (11.3,4,5,7) are called "generalized forces". They are of three different tensorial orders (scalar, vectorial and tensorial) and, as clearly shown by eqs. (11.4,5,7), their actual expression changes as the expressions for the generalized fluxes are changed but they are always expressed in terms of known functions of the set of the "primitive" variables.

Hence the closure of the system of field equations will be obtained when the generalized fluxes are expressed in terms of the generalized forces. Such expressions are called phenomenological relations.

One may say, in principle, that the generalized fluxes depend on the thermidynamic state of the system and on the generalized forces. However, the problem is : how general this dependence may be while remaining within the limits of validity of the expression (11.4) for the entropy production?

For chemical reactions equation (11.4) is compatible with the law of mass action (see later on) (Ref.3). No such general statement is available for the generalized scalar fluxes $\dot{\delta}_j$ although one may speculate, in analogy with what happens with the chemical reactions, that eq. (11.4) must be compatible with functional relationship for the $\dot{\delta}_j$ which are more general than the linear ones (*). For transport processes, the kinetic theory of gases shows that the expression (11.4) holds only as long as linear phenomenological relations (i.e. linear functional dependence between fluxes and forces). Non linear phenomenologies for transport processes (e.g. the theories of non-Newtonian fluids) are being developed which are based either on semi- empirical approaches or on set-theoretical analysis. A discussion of their relationship with the entropy production and of their compatibility with the "shifting-equilibrium" hypothesis falls outside the scope of the present lectures. We shall thus limit ourselves only to linear irreversible thermodynamics. The limits of validity of linear phenomenology are very narrow for chemical reactions. We shall return to this point later on after having briefly exposed the basic postulates of linear irreversible thermodynamics.

## 12. LINEAR IRREVERSIBLE THERMODYNAMICS

We suppose that all fluxes and forces appearing in the several expressions for the production of entropy are independent and shall consider later what modifications must be

---

(*) It has been shown (see Ref. 3) that the law of mass action can be obtained from a "linear" law in some internal coordinate space of the reacting mixture. A similar procedure applied to some internal coordinate space of the vibrational degrees of freedom of the molecules may lead to analogous general expressions for $\dot{s}_i$.

introduced in the case that they are not. As we have seen, the definition of fluxes and forces is by no means unique. In what follows we shall assume as fundamental the definition implied by the following expression for the entropy production:

$$
\begin{aligned}
-T\dot{s} = {} & (\pi - p)\,\underline{\nabla}\cdot\underline{V} + \underline{\underline{\tau}}_0 : (\underline{\nabla}\,\underline{V})_0 + \underline{j}_q' \cdot \frac{\underline{\nabla} T}{T} + \\
& + \sum_{l=1}^{r} A_l \dot{\xi}_l + \sum_{j=1}^{k} (\Theta_j - T)\, s_j + \\
& + \sum_{i=1}^{n-1} \underline{j}_{c_i} \cdot \underline{\nabla} T \,(g_i - g_n) + \sum_{j=1}^{k} \underline{j}_{s_j} \cdot T\, \underline{\nabla}\left(\frac{\Theta_j - T}{T}\right) = \\
= {} & -T\left(\dot{s}_0 + \dot{s}_1 + \dot{s}_2\right) \qquad (12.1)
\end{aligned}
$$

and shall consider later what happens when different definitions are introduced. For simple gas $\underline{j}_q' = \underline{j}_q = T\underline{j}_1 = \underline{j}_u$ and all terms in $\{\Sigma\}$ are absent.

A first basic postulate of linear irreversible thermodynamics is the Curie's postulate which can be formulated as follows:

"In isotropic systems each generalized flux of a given tensorial order depends only upon all generalized forces of the same tensorial order".

Hence:

i) Scalar, vectorial and tensorial fluxes must be linearly related only to scalar, vectorial and tensorial generalized forces, respectively. Thus, for instance, the stress tensor $\underline{\underline{\tau}}_0$ is related to the dyadic $(\underline{\nabla}\,\underline{V})_0$ and not to any gradient of thermodynamic quantities or to quantities such as $(\underline{\nabla}\cdot\underline{V})$, $A_\ell$, $(\Theta_i - T)$.

ii) Each flux depends, in general, upon all the generalized forces of the same tensorial order. This gives rise to so-called cross-coupling phenomena when there are two or more fluxes of the same tensorial order. Thus, for instance, one must expect a coupling between diffusion of mean normal momentum, chemical reactions and relaxation processes. While the coupling between the last two phenomena is considered also in chemical kinetic theories not much is known about the coupling between the first two which entails that, in general, a chemical reaction rate for systems in non-uniform motion should depend also on the divergence of the corresponding mass velocity.Not much is known nor is done, experimentally, to accertain the order of magnitude of this type of coupling.

iii) The coefficients of proportionality between fluxes and forces are, in isotropic media, scalar quantities. They are usually referred to as phenomenological coefficients and are functions of the local thermodynamic state. Quite obviously, macroscopic irreversible thermodynamics is unable to furnish the explicit form of these dependences but, as we shall see presently, can only derive a number of useful inequalities among them.

iv) Since fluxes and forces of different tensorial order are not related, it follows that each of the different contributions $\dot{s}_i$ to the total entropy production must be an essentially positive quantity. Thus:

$$\dot{s}_0 \geq 0 \; ; \quad \dot{s}_1 \geq 0 \; ; \quad \dot{s}_2 \geq 0$$

To simplify the writings we shall employ again matrix notation by letting:

$$\hat{\underline{\varphi}} \equiv \left[ \pi - p \quad \xi_1 ------ \xi_\ell \quad \dot{s}_1 ------ \dot{s}_k \right]$$

$$\hat{\underline{f}} \equiv \left[ \underline{\nabla} \cdot \underline{V} \quad A_1 ----- A_\ell \quad (\Theta_1 - T) ----- (\Theta_k - T) \right]$$

$$\hat{\underline{\Phi}} \equiv \left[ \underline{j}'_q \quad \underline{j}_{c_1} ------ \underline{j}_{c_\ell} \quad \underline{j}_{s_1} ------ \underline{j}_{s_k} \right]$$

$$\hat{\underline{F}} \equiv \left[ \frac{\underline{\nabla}\, T}{T} \quad \underline{\nabla}_T (g_1 - g_n) ----- \underline{\nabla}_T (g_{n-1} - g_n) \quad T \underline{\nabla} \left( \frac{\Theta_1 - T}{T} \right) ----- T \underline{\nabla} \left( \frac{\Theta_k - T}{T} \right) \right]$$

so that equation (12.1) becomes:

$$-T\dot{s} = \underline{\underline{\tau}}_0 : (\underline{\nabla}\, \underline{V})_0 + \tilde{\underline{\varphi}} \cdot \underline{f} + \tilde{\underline{\Phi}} \cdot \underline{F} \tag{12.2}$$

where $(\tilde{\underline{b}})$ is the transpose of $(\underline{b})$ (row matrix) and the dot denotes, as usual, matrix multiplication. Upon the Curie postulate then:

(12.3) $$\underline{\underline{\tau}}_0 = -2\mu(\underline{\nabla}\,\underline{V})_0 \;;\quad \underline{\varphi} = -\underline{\underline{\ell}}\cdot\underline{f} \;;\quad \underline{\Phi} = -\underline{\underline{L}}\cdot\underline{F}$$

where $\underline{\underline{\ell}}$ and $\underline{\underline{L}}$ are the matrices formed by the phenomenological coefficients $\ell_{ij}$, $L_{ij}$ and $\mu$ is the viscosity coefficient. The expression for the entropy production becomes:

(12.4) $$T\dot{s} = 2\mu(\underline{\nabla}\,\underline{V})_0 : (\underline{\nabla}\,\underline{V})_0 + \tilde{\underline{f}}\cdot\underline{\underline{\ell}}\cdot\underline{f} + \tilde{\underline{F}}\cdot\underline{\underline{L}}\cdot\underline{F} > 0 \,.$$

Since the forces have been assumed to be independent, it follows that $\mu > 0$ and that the matrices $\underline{\underline{\ell}}$ and $\underline{\underline{L}}$ are positive definite. Hence the diagonal elements of these matrices are all essentially positive: they are the "direct" phenomenological coefficients: they may be either positive or negative and satisfy a number of inequalities deriving from the condition that all principal minors of the matrices $\underline{\underline{\ell}}$ and $\underline{\underline{L}}$ are positive. The positive quantity $2\mu(\underline{\nabla}\,\underline{V})_0 : (\underline{\nabla}\,\underline{V})_0$ is often referred to in fluid-dynamic as the "dissipation fuction".

Before enunciating the second basic postulate it proves useful to introduce the notion of compatible transformations of fluxes and forces. The reasoning is general and applies to fluxes and forces of any tensorial character. We shall carry it on only for vectorial fluxes.

Let:

(12.5) $$T\dot{s}_1 = -\tilde{\underline{\Phi}}\cdot\underline{F} = \underline{F}\;\underline{\underline{L}}\;\underline{F}$$

be the expression for the entropy production associated with the vectorial fluxes and consequent to the phenomenological relation:

$$\underline{\Phi} = -\underline{\underline{L}} \cdot \underline{F} \qquad (12.6)$$

Let a linear transformation of fluxes and forces be defined by:

$$\underline{\Phi}_1 = \underline{\underline{X}} \cdot \underline{\Phi} \qquad \underline{F}_1 = \underline{\underline{Y}} \cdot \underline{F} \qquad (12.7)$$

where the square matrices $\underline{\underline{X}}$ and $\underline{\underline{Y}}$ are arbitrary functions of the thermodynamic state of medium. Suppose that the matrix $\underline{\underline{Y}}$ is non-singular so that also the new forces are independent. Such a transformation will be said compatible if it leaves unchanged the expression for the entropy production, i.e. if:

$$T\dot{s} = -\underline{\tilde{\Phi}} \cdot \underline{F} = -\underline{\tilde{\Phi}}_1 \cdot \underline{F}_1 \qquad (12.8)$$

Substitution of equations (12.7) into this equation gives:

$$\underline{\tilde{\Phi}} \cdot \underline{F} = \underline{\tilde{\Phi}} \cdot \underline{\underline{\tilde{X}}} \cdot \underline{\underline{Y}} \cdot \underline{F} \qquad (12.9)$$

as the condition defining a compatible transformation. The general solution of this equation is (Ref. 5):

$$\underline{\underline{\tilde{X}}} \cdot \underline{\underline{Y}} = \underline{\underline{U}} + \underline{\underline{A}} \cdot \underline{\underline{L}} \qquad (12.10)$$

where $\underline{\underline{U}}$ is the unit matrix and $\underline{\underline{A}}$ an arbitrary antisymmetric matrix [i.e. $\underline{\underline{\tilde{A}}} = -\underline{\underline{A}}$] Indeed when equation (12.10) is substituted into eq. (12.9) and account is taken of the phenomenological relation (12.6) one gets:

$$\tilde{\underline{\Phi}} \cdot \tilde{\underline{\underline{X}}} \cdot \underline{\underline{Y}} \cdot \underline{F} = \tilde{\underline{\Phi}} \cdot \underline{F} + \tilde{\underline{\Phi}} \cdot \underline{\underline{A}} \cdot \underline{\underline{L}} \cdot \underline{F} =$$

$$= \tilde{\underline{\Phi}} \cdot \underline{F} - \tilde{\underline{\Phi}} \cdot \underline{\underline{A}} \cdot \underline{\Phi} = \tilde{\underline{\Phi}} \cdot \underline{F}$$

since the term $\tilde{\underline{\Phi}} \cdot \underline{\underline{A}} \cdot \underline{\Phi}$ vanishes identically, due to the antisymmetry of $\underline{\underline{A}}$ .

In terms of the news fluxes and forces the phenomenological relation becomes, upon substitution of eqs. (12.7) into eq. (12.6):

$$\underline{\Phi}_1 = - \underline{\underline{X}} \cdot \underline{\underline{L}} \cdot \underline{\underline{Y}}^{-1} \cdot \underline{F}_1 = - \underline{\underline{L}}_1 \cdot \underline{F}_1$$

where

$$\underline{\underline{L}}_1 = \underline{\underline{X}} \cdot \underline{\underline{L}} \cdot \underline{\underline{Y}}^{-1} \tag{12.11}$$

is the new phenomenological matrix. For a <u>compatible</u> transformation, the phenomenological <u>matrix</u> becomes:

$$\underline{\underline{L}}_1 = \underline{\underline{X}} \cdot \left[ \underline{\underline{L}}^{-1} + \underline{\underline{A}} \right]^{-1} \cdot \tilde{\underline{\underline{X}}} \tag{12.12}$$

We can now state the Onsager postulate which will be given in its "restricted" form since we are not interested here in phenomena involving the presence of imposed external magnetic fields or of Coriolis forces (i.e. rotating system). This restricted form is as follows:

"It is always possible to define a set of independent generalized forces and fluxes such that the corresponding phenomenological matrix is symmetric".

To understand the full implication of this postulate notice that:

- if for a given set of independent forces and fluxes the phenomenological matrix is symmetric this will not be necessarily so for any compatible transformation of forces and fluxes. Indeed, as equation (12.12) clearly shows, if $\underline{\underline{L}}$ is symmetric $\underline{\underline{L}}_1$ will not be symmetric unless $\underline{\underline{A}}$ is the null matrix. Hence, given a set of forces and fluxes for which the symmetry relation is valid, all other sets for which the same relation applies are defined by the following particular class of compatible transformations [see eq. (12.10) with $\underline{\underline{A}} = 0$]

$$\underline{\Phi}_1 = \underline{\underline{\tilde{Y}}}^{-1} \cdot \underline{\Phi} \qquad \underline{F}_1 = \underline{\underline{Y}} \cdot \underline{F} \tag{12.13a}$$

or, equivalently:

$$\underline{\Phi}_1 = \underline{\underline{X}} \cdot \underline{\Phi} \qquad \underline{F}_1 = \underline{\underline{\tilde{X}}}^{-1} \cdot \underline{F} \tag{12.13b}$$

and the relationship between the two phenomenological matrices $\underline{\underline{L}}$ and $\underline{\underline{L}}_1$ is:

$$\underline{\underline{L}}_1 = \underline{\underline{X}} \cdot \underline{\underline{L}}^{-1} \cdot \underline{\underline{\tilde{X}}} \tag{12.14}$$

In other words: if the other transformation must leave invariant the expression for the entropy production and the symmetric property of the kinetic matrices <u>only the forces (or the fluxes) may be transformed independently.</u> Notice also that $\underline{\underline{X}}$ is non-sin-gular, i.e. <u>the new fluxes are also independent</u>.

By reversing the reasoning just made, one finds that if for a given set of forces and fluxes the phenomenological matrix is not symmetric one can always determine compatible transformations leading to new sets of forces and fluxes for which the phenomenological matrices are symmetric (*).

Hence the postulate is misstated when the symmetry of the phenomenological coefficients is ascribed to either <u>any set</u> of generalized fluxes and forces or to <u>none</u>.

We shall "assume" that the Onsager postulate holds for the system of fluxes and forces defined by eq. (12.1). It can be readily proved [see Example 16] that it then holds also for the system defined by equations (11.4) and (11.5).

---

(+) Suppose indeed, that L is not symmetric. Let $\underline{\underline{B}} = \underline{\underline{L}}^{-1}$ and denote with $\underline{\underline{B}}^s$ and $\underline{\underline{B}}^a$ its symmetric and antisymmetric part. Then the condition that $\underline{\underline{L}}$ be symmetric leads to the matricial equation:

$$\underline{\underline{B}} - \underline{\underline{A}} = \tilde{\underline{\underline{B}}} - \underline{\underline{A}}$$

whose solution is $\acute{\underline{\underline{A}}} = (\hat{\underline{\underline{B}}} - \underline{\underline{B}})/2 = -\underline{\underline{B}}^a$. Hence eq. 12.10 becomes

$$\tilde{\underline{\underline{X}}} \cdot \underline{\underline{Y}} = (\underline{\underline{B}} + \underline{\underline{A}}) \cdot \underline{\underline{L}} = \underline{\underline{B}}^s \cdot \underline{\underline{L}} = (L^{-1})^s \cdot \underline{\underline{L}}$$

and any transformation of the type:

$$\Phi_1 = \underline{\underline{X}} \cdot \Phi \qquad \underline{F}_1 = \tilde{\underline{\underline{X}}}^{-1} \cdot (\underline{\underline{L}}^{-1})^s \cdot \underline{\underline{L}} \cdot \underline{F}$$

will lead to a symmetric $\underline{\underline{L}}_1$. Notice that when $\underline{\underline{L}}$ is symmetric the second of these equations reduces to eqs. $(12.14b)_2$.

Example 16

For the "typical" system there are 3 generalized forces of vectorial order. With the expression for the entropy production given by eq. (12.1) the new matrices $\tilde{\underline{F}}$ and $\tilde{\underline{\Phi}}$ are

$$\tilde{\underline{F}} \equiv \left[ \frac{\underline{\nabla} T}{T} \quad \underline{\nabla}_T(g_1 - g_2) \quad T\underline{\nabla}\left(\frac{\Theta - T}{T}\right) \right]$$

$$\tilde{\underline{\Phi}} \equiv \left[ \underline{j}'_q \quad \underline{j}_c \quad \underline{j}_{sv} \right]$$

On the other hand, with the expression for the entropy production given by eq. (11.5) the row matrix $\tilde{\underline{F}}_1$, is:

$$\tilde{\underline{F}}_1 \equiv \left[ \frac{\underline{\nabla} T}{T} \quad T\underline{\nabla}\left(\frac{g_1 - g_2}{T}\right) \quad T\underline{\nabla}\left(\frac{\Theta - T}{T}\right) \right]$$

Hence, by recalling that eq. 5.12

$$T\underline{\nabla}\left(\frac{g_i}{T}\right) = \underline{\nabla}_T g_i - \hat{h}_i \frac{\underline{\nabla} T}{T}$$

one finds that the transformation matrix $\underline{\underline{Y}}$ is given by:

$$\underline{\underline{Y}} = \begin{bmatrix} 1 & 0 & 0 \\ \hat{h}_2 - \hat{h}_1 & 1 & 0 \\ 0 & 0 & 1 \end{bmatrix}$$

Thus:

$$\underset{=}{\tilde{Y}}^{-1} = \begin{bmatrix} 1 & \hat{\underset{\sim}{h}}_1 - \hat{\underset{\sim}{h}}_2 & 0 \\ 0 & 1 & 0 \\ 0 & 0 & 1 \end{bmatrix}$$

and the transformed fluxes which preserve the Onsager's symmetry relations are given by [see eq. (12.14a)]:

$$\Phi_1 = \underset{=}{\tilde{Y}} \cdot \underline{\Phi} = \begin{bmatrix} \underline{j}'_q + (\hat{\underset{\sim}{h}}_1 - \hat{\underset{\sim}{h}}_2)\underline{j}_c \\ \underline{j}_c \\ \underline{j}_{sv} \end{bmatrix}$$

But [see eq. (10.9) with $\underline{j}_{c_1} = \underline{j}_c$ ; $\underline{j}_{c_2} = -\underline{j}_c$]:

$$\underline{j}'_q + \hat{\underset{\sim}{h}}_1 \underline{j}_{c_1} + \hat{\underset{\sim}{h}}_2 \underline{j}_{c_2} = \underline{j}'_q + (\hat{\underset{\sim}{h}}_1 - \hat{\underset{\sim}{h}}_2)\underline{j}_c = \underline{j}_q$$

The proof carries over, unchanged, to the case of an arbitrary number of $g_i$. Hence: the Onsager's postulate holds also for the set of fluxes and forces defined by equation (11.5).

Turning now to the set of vectorial generalized forces defined by eq. (11.4) one has:

$$\tilde{\underline{F}}_1 = \left[ \underline{\nabla} T \quad \underline{\nabla}(g_1 - g_2) \quad \underline{\nabla}(\Theta - T) \right]$$

Then, since, upon eqs. (5.12) and (5.11):

$$\underline{\nabla} g_i = \underline{\nabla}_T g_i + \frac{1}{T}\left(g_i - \hat{\underset{\sim}{h}}_i\right) \underline{\nabla} T = \underline{\nabla}_T g_i - \hat{s}_i \underline{\nabla} T$$

the transformation matrix reads:

$$\underline{\underline{Y}} = \begin{bmatrix} T & 0 & 0 \\ (\hat{s}_2 - \hat{s}_1)T & 1 & 0 \\ (\Theta - T) & 0 & 1 \end{bmatrix}$$

and proceeding as before one finds that the fluxes preserving the Onsager relations are :

$$\frac{1}{T}\, \underline{j}'_q + \left(\hat{s}_1 - \hat{s}_2\right) \underline{j}_c - \frac{\Theta - T}{T}\, \underline{j}_{sv} \; ; \; \underline{j}_c \; ; \; \underline{j}_{sv}$$

Comparison with eq. $(10.10)_2$ proves that the Onsager's postulate holds also for the set of fluxes and forces defined by eq. (11.4).

Example 17

As discussed in section (9) when the (n-1) balance equations for the masses $M_i$ are replaced by r balance equations in terms of the progress variables $\xi_\ell$ plus $(n-1-r) = m-1$

$[m = n - r]$ conservation equations for the variables $\sigma_i$ the set of a mass diffusion fluxes is replaced by the $n$ fluxes $\underline{j}_{\xi_\ell}, (1 \le \ell \le r)$ and $\underline{j}_{\sigma_i}, (1 \le i \le m)$ defined by equations (8.10). To simplify the writings we shall for a moment consider all the $(n)$ diffusion fluxes although we know that only $(n-1)$ of them are independent. Hence, if

$$\tilde{\underline{\Phi}} \equiv \left[ \underline{j}_{c_1} \text{------} \underline{j}_{c_r} \quad \underline{j}_{c_{r+1}} \text{------} \underline{j}_{c_{r+m}} \right]$$

$$\tilde{\underline{F}} \equiv \left[ \underline{\nabla}_T g_1 \quad \underline{\nabla}_T g_r \quad \underline{\nabla}_T g_{r+1} \quad \underline{\nabla}_T g_{r+m} \right]$$

the introduction of the new fluxes:

$$\underline{\Phi}_1 = \left[ \underline{j}_{\xi_1} \text{------} \underline{j}_{\xi_r} \quad \underline{j}_{\sigma_1} \text{------} \underline{j}_{\sigma_m} \right]$$

implies the following transformation (see eqs. 8.8 and 8.10)

$$\underline{\Phi}_1 = \begin{vmatrix} \tilde{\underline{\underline{R}}}^{-1} & \underline{\underline{0}} \\ -\tilde{\underline{\underline{Q}}} \cdot \tilde{\underline{\underline{R}}}^{-1} & \underline{\underline{U}} \end{vmatrix} \cdot \underline{\Phi} = \underline{\underline{M}} \cdot \underline{\Phi}$$

We would like to know what is the set of new generalized forces $\underline{F}_1$ for which the Onsager's relations hold. From equation (12.14b) one gets (with $\underline{\underline{X}} \equiv \underline{\underline{M}}$):

$$\underline{F}_1 = \tilde{\underline{\underline{M}}}^{-1} \cdot \underline{F}$$

or, [see eq. 8.11]

$$\underline{F}_1 = \begin{vmatrix} \underline{\underline{R}} & \underline{\underline{Q}} \\ \underline{\underline{0}} & \underline{\underline{U}} \end{vmatrix} \cdot \underline{F}$$

But (see eq.8.3) the matrix $|\underline{\underline{R}} \quad \underline{\underline{Q}}| \equiv \underline{\underline{N}}$ where $\underline{\underline{N}}$ is the $(r \times n)$ matrix of the stoichiometric coefficients defined by eq. (8.2). Hence the first $r$ elements of $\underline{F}_1$ are given by:

$$\sum_{i=1}^{n} \nu_{\ell i} \, \underline{\nabla}_T g_i = \underline{\nabla}_T \left( \sum_{i=1}^{n} \nu_{\ell i} \, g_i \right) = \underline{\nabla}_T A_\ell \qquad (1 \le \ell \le r)$$

where the $A_\ell$ are the de-Donder affinities of the $r$ independent reactions. The remaining $m = n - r$ elements of $\underline{F}_1$ coincide with the corresponding elements of $\underline{F}$ . Thus the contribution to the entropy production due to mass diffusion fluxes:

$$\sum_{i=1}^{n-1} \underline{j}_{c_i} \cdot \underline{\nabla}_T (g_i - g_n)$$

can also be written as:

$$\sum_{\ell=1}^{r} \underline{j}_{\xi_\ell} \cdot \underline{\nabla}_T A_\ell + \sum_{j=1}^{m-1} \underline{j}_{\sigma_j} \cdot \underline{\nabla}_T (g_j - g_n) \qquad (n = r + m)$$

and in either case the Onsager's relations apply.

We have so far supposed that the fluxes and forces were all independent and the formulation given here to the Onsager's postulate is explicitly based on the hypothesis. It may be of interest to investigate about the validity of the Onsager's relations when such is no longer the case.

Three possibilities may arise:

1) The forces are not independent (*)
2) The fluxes are not independent
3) Both forces and fluxes are not independent.

We shall only state the results (see Ref. 3 for details and proofs).

1) The Onsager's postulate cannot be applied when the forces are not independent. Thus before invoking the Onsager's symmetry relations one must be sure that the forces are independent. In the other two cases the Onsager's postulate applies. More rigorously (Ref.5):
2) A linear homogenous relationship among the fluxes does not impair the validity of the Onsager's postulate.
3) When linear homogenous relationships exist among both the

---

(*) It may be appropriate to mention that upon the Gibbs-Duhem relation it is:

$$s \underline{\nabla} T - v \underline{\nabla} p + \sum_{i=1}^{n-1} e_i \underline{\nabla}(g_i - g_n) + \sum_{j=1}^{k} s_j \underline{\nabla}(\theta_j - T) = 0.$$

Hence the vectorial generalized forces appearing in eq. (11.3) [and consequently, in all other expressions for $\dot{s}$ ] are independent only if $\underline{\nabla} P \neq 0$, that is only when the so called "mechanical equilibrium" does not prevail.

fluxes and the forces the phenomenological coefficients are not uniquely defined but they can always be chosen so as to satisfy the Onsager's postulate.

To discuss the relevant consequences of the basic postulates of linear irreversible thermodynamics in a simple manner we shall deal almost always with case, already analysed from the equilibrium thermodynamic point of view, of an (ideal) vibrating and dissociating biatomic gas.

We shall consider at some length only the scalar and vectorial fluxes. The discussion concerning the tensorial flux is completely similar to that relative to a pure gas and the only matter which should be further discussed is the nature of the dependence of the viscosity coefficient $\mu$ upon the state parameters of the systems. As often mentioned, this cannot be done within the framework of a macroscopic theory and statistical theories and/or experiments must be resorted to.

## 13. SCALAR FLUXES. CHEMICAL KINETICS OF THE RATE PROCESSES

For the scalar fluxes we shall first discuss briefly the linear phenomenological relations. We shall then examine the determination of the rates for the two relevant processes (dissociation and vibrational excitation) from the kinetics point of view both to illustrate the narrow limits of validity of linear phenomenologies for the reaction rates and to obtain the explicit expressions for the phenomenological coefficients.

## 1. Linear phenomenology

In the case of a biatomic dissociating and vibrating gas there are three scalar fluxes which, with the notation of section (6), are:

$(\pi - p)$, $\dot{C}$, $\dot{s}_v$. The corresponding conjugate forces are:

$(\underline{\nabla} \cdot \underline{V})$; $A = g_1 - g_2$; $(\Theta - T)$. The phenomenological relations read:

$$
\begin{aligned}
-\dot{C} &= \ell_{11} A + \ell_{12} (\Theta - T) + \ell_{13} \underline{\nabla} \cdot \underline{V} \\
-\dot{s}_v &= \ell_{21} A + \ell_{22} (\Theta - T) + \ell_{23} \underline{\nabla} \cdot \underline{V} \qquad (13.1) \\
-(\pi - p) &= \ell_{31} A + \ell_{32} (\Theta - T) + \ell_{33} \underline{\nabla} \cdot \underline{V}
\end{aligned}
$$

and the positivity of entropy production implies that:

$$
\ell_{ii} > 0 \; ; \quad \ell_{ii}\ell_{jj} - \ell_{ij}\ell_{ji} > 0 \qquad \det \underline{\underline{\ell}} > 0 \qquad (i,j = 1,2,3) \qquad (13.2)
$$

The direct coefficient $\ell_{33}$ represents the well known bulk viscosity coefficient. The Onsager reciprocity relations imply that $\ell_{ij} = \ell_{ji}$. The coefficients $\ell_{3j} = \ell_{j3}$ $(j = 1,2)$ give a measure of the cross coupling between mean normal stress and the rate processes. The coefficient $\ell_{12}$ is a measure of the kinetic coupling between the two rate processes. The phenomenological relations (13.1) show that the condition necessary for the prevailing of partial equilibria is that the two direct coefficients $\ell_{11}$ and $\ell_{22}$ be of different order of magnitude. This condition is also sufficient if coupling is negligible. As we shall see presently the limits of valifity of a linear phenomenology for the reaction rate $\dot{C}$ are very narrow. The linear approach, however, brings out two important points. The first one is related to the appearance of the terms in $\underline{\nabla}\cdot\underline{V}$ in equation (13.1). This casts some doubts on the validity of the straight-forward application of results obtained from chemical kinetics (which are almost always based on the hypothesis of uniformity) to the non-uniform evolution of reacting mixtures. Not much is known on the relative order of magnitude of these effects except what implied by the second inequality (13.2) which read, in this case:

$$0 < \frac{\ell_{13}^2}{\ell_{11}\,\ell_{33}} < 1 \ ; \quad 0 < \frac{\ell_{23}^2}{\ell_{22}\,\ell_{33}} < 1$$

The second point is related to the coupling between the two rate processes, as measured by the coefficient $\ell_{12}$ appearing in eqs. (13.1). It indicates that an adequate description of the two rate processes from the kinetics point of view must necessarily account for the presence of this coupling.

We shall discuss the kinetics approach in the next two paragraphs. The first one is somewhat introductory and is meant to bring forth two points: a) the inadequacy of reactions rates based on the straight-forward extension of the law of mass action when there is thermal non equilibrium, b) the validity of a linear phenomenology for chemical reaction is restricted by the condition that $(\bar{A}_j / R_0 T) \ll 1$ where $\bar{A}_j$ is the affinity per mole. This will be done, for the sake of simplicity, for the case of a single reaction.

In the second paragraph a more sophisticated model of the two rate processes for a biatomic dissociating and vibrating gas will be derived and discussed. This model accounts for the coupling between the two rates. It will be used to obtain the explicit expressions for the phenomenological coefficients $\ell_{ij}$ $(i,j=1,2)$ appearing in the phenomenological relations. In both cases the coupling with the diffusive flux of normal momentum is not included: the treatment of chemical reactions in non-uniform media, either alone or in the presence of other rate processes, has not yet reached a satisfactory state.

## 2. Chemical reaction according to the law of mass action.

For simplicity's sake we shall consider that only one reaction:

$$(13.3) \qquad \sum_{i=1}^{n} \bar{\nu}_i \left[C_i\right] \rightleftharpoons \sum_{i=1}^{n} \bar{\nu}'_i \left[C_i\right]$$

takes place in an ideal mixture of perfect gases.

We shall assume that the law of mass action holds i.e. that the reaction does not appreciably destroy the equilibrium Maxwell distribution non-equilibrium. This law is usually stated in terms of mole densities $n_i = (\varrho_i / m_i)$ where $\varrho_i$ is the density and $m_i$ the molar mass of the $i$-th constituent.

Let:

$$\Delta \bar{\nu}_i = (\bar{\nu}_i' - \bar{\nu}_i) \tag{13.4}$$

and denote by $K_f$ and $K_b$ the (positive) rate constants for the forward $(\rightarrow)$ and backward $(\leftarrow)$ reactions, respectively.

The production $\dot{C}_i$ (mass of the $i$-th element per unit volume and time) can then be written, on account of eq. (2.2) as (*):

$$\dot{C}_i = (\Delta \bar{\nu}_i) m_i k_b \prod_{i=1}^{n} n_i^{\bar{\nu}_i'} \left[ \frac{K_f}{K_b} \prod_{i=1}^{n} n_i^{-\Delta \bar{\nu}_i} - 1 \right] \tag{13.5}$$

(*) Notice that the law of mass action, as usually formulated, presupposes that the specific volume remains constant. Thus strictly speaking it should be used for a suitable fiducial specific volume $v_0$. Consequently, in the expression for $\dot{C}_i$ one should add the contribution due to the substantial rate of change of volume (since the reference frame in which the $\dot{C}_i$ is computed is the one moving with the mass velocity $\underline{V}$ ). This contribution as easily shown, amounts, on account of the continuity equation, to $-\varrho_i \underline{\nabla} \cdot \underline{V}$.

This equation defines the ratio $(K_f/K_b)$ in terms of the values assumed by the quantity $\left(\prod_{j=1}^{n} n_j^{-\Delta\bar{\nu}_j}\right)$ for $\dot{C} \equiv 0$. For a mixture in thermal equilibrium [but <u>only</u> for it, unless the coupling is negligible] the condition $\dot{C} \equiv 0$ is equivalent to the condition $A = 0$ (where $A$ is the affinity of the reaction) and one can therefore evaluate $(K_f/K_b)$ from the "equilibrium thermodynamic" description of the mixture for then, by definition:

$$K = \frac{K_f}{K_b} = \prod_{j=1}^{n} (n_{je})^{-\Delta\bar{\nu}_j} \tag{13.6}$$

where $K$ is the so-called equilibrium "constant" and the $n_{je}$ are the "equilibrium" values of the mole densities.

As discussed in section (6) the electrochemical potential per mole $\mu_i = m_i g_i$ for an ideal mixture of perfect gases in thermal equilibrium can be written as:

$$\mu_j = \mu_{jo}(T) + R_0 T \ln \frac{n_j}{n_0} \tag{13.7}$$

where $n_0$ is a fiducial mole density and $\mu_{jo}$ is the electrochemical potential per mole of the "pure" gas $(j)$.

Substituting into eq. (13.5) the molar densities $n_j$ as obtained from eq. (13.7) one gets:

$$\dot{C}_i = (\Delta\bar{\nu}_i) m_i K_b \prod_{j=1}^{n} n_j^{\bar{\nu}_j} \left[ \frac{K_f}{K_b} n_0^{-\Delta\bar{\nu}} \exp\left( \frac{\sum_{j=1}^{n} \Delta\bar{\nu}_j \mu_{jo} - \bar{A}}{R_0 T} \right) - 1 \right] \tag{13.8}$$

where:

$$\Delta\bar{\nu} = \sum_{j=1}^{n} \Delta\bar{\nu}_j$$

is the total change in mole numbers induced by the reaction and see eqs. (2.2) and (2.14) :

$$\bar{A} = \sum_{j=1}^{n} (\bar{\nu}_j' - \bar{\nu}_j)\mu_j = \sum_{j=1}^{n} \nu_j g_j m_r = A m_r \tag{13.9}$$

$\bar{A}$ is the affinity (per mole) of the reaction; $m_r$ a reference molar mass and $A$ the affinity as defined in eq. (2.14).

In condition of chemical equilibrium $A=0, \dot{C}_i=0$ as said. Hence:

$$K = \frac{K_f}{K_b} = n_0^{\Delta\bar{\nu}} \exp\left[-\sum_{j=1}^{n} \frac{\Delta\bar{\nu}_i \mu_{jo}}{R_0 T}\right] = K(T) \tag{13.10}$$

so that eq.(13.8) can also be written as :

$$\dot{C}_i = \Delta\bar{\nu}_i m_i K_b \prod_{j=1}^{n} n_j^{\bar{\nu}_j'} \left[\exp\left(-\frac{\bar{A}}{R_0 T}\right) - 1\right]$$

and the rate of change $\dot{\xi} = \dot{C}_i / \nu_i = (C_i m_r)/(\Delta\bar{\nu}_i m_i)$ of the progress variable [see eqs. (2.2) and (2.11)] reads:

$$\dot{\xi} = m_r K_b \prod_{j=1}^{n} n_j^{\bar{\nu}_j'} \left[\exp\left(-\frac{\bar{A}}{R_0 T}\right) - 1\right] \tag{13.11}$$

The limits of validity of a linear phenomenology

for the chemical rates are seen to be defined by the condition

$$\left(\bar{A} / R_0 T\right) \ll 1$$

Indeed, by developing eq. (13.11) around the "point" $A = 0$ one obtains, to within terms of order $\left(\bar{A} / R_0 T\right)$:

$$\dot{\xi} = - m_r K_b \prod_{j=1}^{n} \left(n_{je}\right)^{\bar{\nu}_i'} \frac{\bar{A}}{R_0 T}$$

a relation which serves to identify the linear phenomenological coefficient. We also clearly see why the present approach cannot account for any coupling between chemical reactions and other rate processes. In order to arrive at the expression (13.6) we <u>had</u> to assume that couplings were negligible (otherwise $\dot{C}_i = 0$ would not imply $A = 0$). Thus, even if we were to use for the electrochemical potential $\mu_j$ the more general relation:

$$\mu_j = \mu_{jo}(T, \Theta) + R_0 T \ln \frac{n_j}{n_0}$$

valid (upon the shifting equilibrium hypothesis) for ideal mixtures in thermal non-equilibrium only the functional expression for $K$ would change [it would now depend on both $T$ and $\Theta_r$]. No "coupling" term would be present in the expression for $\dot{\xi}$ which would still be given by eq. (13.11) [with, of course, $\bar{A}$ now function also of $\Theta_r$].

An altogether more sophisticated approach is therefore needed to account for kinetic coupling. This will be exposed in the next paragraph.

## 3. Kinetic model of coupled reaction and vibrational excitation rates.

In this paragraph we shall present the kinetic models of two coupled rate processes such as they occur in the system formed by an ideal biatomic dissociating and vibrating gas. The "equilibrium thermodynamic" description of this model has already been discussed at length. The treatment of its rate processes will complete the description of this system also from the irreversible thermodynamic point of view.

The subject model was originally presented in (Refs. 5,8) and further elaborated upon in (Ref.8). The presentation here will follow very closely that of (Ref. 8).

The kinetic model is based on the "harmonic oscillator with cut-off" model for vibrators [see paragraph 6.4]

### A. Reaction rate -

The dissociation reaction is written as:

$$[X_2]+[M] \underset{K_b}{\overset{K_f}{\rightleftharpoons}} 2[X]+[M] \qquad (13.12)$$

where $[M]$ denotes the "third body".

The molar densities of the atoms $(n_1)$ and molecules $(n_2)$ are related to density $\varrho$ and the atom's mass concentration $c$ by:

$$n_1 = \frac{2c}{\upsilon m_2} \qquad n_2 = \frac{1-c}{\upsilon m_2} \tag{13.13}$$

where $m_2$ is the molar mass of the molecules. Furthermore:

$$\Delta \bar{V}_1 = 2 \qquad \Delta \bar{V}_2 = -1 \qquad m_2 = 2m \quad \text{and} \quad \Delta \bar{V}_3 = 0$$

[where the index (3) relates to the third body (M)]

We write again for $\dot{c} = \dot{c}_1$ the expression:

$$\dot{c} = 2m_1 n_3 \left[K_f n_2 - K_b n_1^2\right] =$$

$$= m_2 n_3 \left[K_f \frac{(1-c)}{m_2 V} - K_b \frac{4C^2}{V^2 m_2^2}\right] = \tag{13.14}$$

$$= \frac{4 n_3 K_b}{V_0 m_2 V}\left[K_1(1-c) - \frac{V_0 c^2}{V^2}\right]$$

with $n_3$ the molar concentration of the third body and

$$K_1 = \frac{m_2 V_0}{4} \; \frac{K_f}{K_b}$$

The non-dimensional quantity $K_1$ is, by definition:

$$K_1 = \left[\left(\frac{V_0}{V}\right)\frac{c^2}{1-c}\right]_{\dot{c}=0}$$

and, as discussed in the previous paragraph, we cannot assume that $\dot{c}=0$ implies $A=0$ when there is coupling. We know, however, that its equilibrium value $K_{1c}$:

$$K_{1c} = (K_1)_{\dot{S}V=0} = (K_1)_{A=0},\ \Theta = T$$

has a well defined functional dependence on the temperature, obtainable from the assumed thermodynamic model. Indeed from eq. (6.25) with $A=0, T=\Theta$ one gets:

$$K_{1c} = \left[\frac{c^2}{1-c}\,\frac{V_0}{V}\right]_{A=0,\Theta=T} =$$

$$= e^{-(D_0/R_0T)}\left(\frac{T}{T_0}\right)^{\frac{1}{2}} Q^{-1}(T) \qquad (13.15)$$

with $Q(T)$ given by eq. (6.23).

If the subscript $(V_e)$ denotes values computed in conditions of vibrational equilibrium we put, in general:

$$K_1 = \frac{V_0 m_2}{4}\;\frac{K_f}{(K_f)\,Ve}\;\frac{(K_b)\,Ve}{K_b}\;\frac{(K_f)\,Ve}{K_bVe} = \frac{K_f}{(K_f)Ve}\;\frac{(K_b)Ve}{K_b}\;K_{1e} \qquad (13.16)$$

since, as just seen, when $\Theta=(K_1)$ coincides with the thermodynamically determined function $K_{1e}(T)$.

The ratio $\left[(K_b)Ve/K_b\right]$ measures the effect of the thermal non-equilibrium on the recombination rate. In three body encounters involving two atoms and a molecule the vibrational excitation of the molecule cannot play an important role as far as the recombination is concerned. In other words the affectiveness of the recombination should not appreciably depend on the vibrational excitation of the third body so that it appears reasonable to put:

$$K_b = (K_b)\,Ve \tag{13.17}$$

The only difference between $K_1$ and $K_{1e}$ can thus be ascribed to the influence of vibrational non-equilibrium excitation on the forward (dissociation) rate.

If one assumes that $K_f$ is proportional to the probability that a molecule be in a given vibrational energy level $(E_v)$ multiplied by the number of molecules having translation energies equal to or greater than the difference between the dissociation energy of the molecule and $E_v$ one may write [see Ref. 7]:

$$K_f \simeq \exp\left(-\frac{D_0}{R_0 T}\right)\cdot\frac{Q(\Theta_m)}{\Theta(\Theta)} \tag{13.18}$$

where $Q(\Theta_m)$ is the partition function given by eq. (6.23) computed at the "temperature" $\Theta_m$ defined by:

$$\frac{1}{\Theta_m} = \frac{1}{\Theta} - \frac{1}{T}$$

Since at thermal equilibrium $\Theta_m \to \infty$ and (see eq. 6.23):

$$\lim_{\Theta \to \infty} Q(\Theta) = \frac{D}{R_0\bar{\Theta}} = \frac{\bar{T}_D}{\bar{\Theta}}$$

we obtain, from eq. (13.18):

$$(K_f)\,Ve = \exp\left[-\frac{D_0}{R_0T}\right]\cdot\frac{D_0}{R_0\bar{\Theta}}\,Q^{-1}(T)$$

$$\frac{K_f}{(K_f)\,Ve} = \frac{R_0\bar{\Theta}}{D_0}\,\frac{Q(\Theta_m)\,Q(T)}{Q(\Theta)} \tag{13.19}$$

Thus, combining eqs. (13.15), (13.16), (13.17) and (13.19) one gets the required expression for $K_1$ as:

$$K_1 = \frac{R_0\bar{\Theta}}{D_0}\left(\frac{T}{T_0}\right)^{\frac{1}{2}}\frac{Q(\Theta_m)}{Q(\Theta)}\exp\left[-\frac{D_0}{R_0T}\right] \tag{13.20}$$

This relation should be compared with the one obtained from the thermodynamic model discussed in paragraph (6.4) under the hypothesis that $K_1 \equiv (K_1)_{A=0}$. From eq. (6.25) one gets:

$$(K_1)_{A=0} = \left[\frac{c^2}{1-c}\left(\frac{V_0}{V}\right)\right]_{A=0} = \left(\frac{T}{T_0}\right)^{\frac{1}{2}}\left[Q(\Theta)\right]^{\frac{\Theta}{T}}\exp\left[-\frac{D_0}{R_0T}\right]$$

Substitution of eq. (13.20) into eq. (13.14) leads, when accounting for eq. (13.17), to the required expression for $\dot{c}$ . Everything is here known except $(K_b)V_0$ whose explicit expression must be found from experiments.

B. Vibrational excitation rate -

As fully discussed in Ref. (4) the equation expressing the time rate of change of the average energy associated with the vibration of the molecules must also take into account the energy lost by the vibrators because of the dissociation and the energy gained because of the recombination (these terms give the coupling between the two rate processes). From this energy rate equation one can readily obtain the expression for the production $\dot{S}_v$ [see Ref. 7].

With the present notation one has:

$$\dot{S}_v = \frac{1}{t_v}\,\frac{1-c}{\theta}\left[E(t)-E(\theta)\right]-$$

$$-\dot{c}_t\left[\frac{m_2 E(\theta_m)}{D_0}+\ln Q\left(\theta_m\right)\right]+\dot{c}_b\left[\frac{m_2 E(\infty)}{D_0}+\ln Q(\theta)\right]$$

where $\dot{c}_t$ and $\dot{c}_b$ are the production of $c$ due to the forward and backward reaction, respectively; $t_v$ is a characteristic time associated with the termalization process, and the function is given by:

$$E(y) = \frac{R_0}{m_0}\left\{\frac{\bar{\theta}}{\exp(\theta/y)-1}-\frac{(D_0/R_0)}{\exp\left(\frac{D_0}{R_0 y}\right)-1}\right\}$$

so that :

$$\frac{m_2 E(\infty)}{D_0} = \frac{1}{2}\left[1-\frac{\bar{\theta}R_0}{D_0}\right]$$

C. <u>Linearized rate equations</u> -

The phenomenological coefficients appearing in the linear phenomenology:

$$-\dot{c} = \ell_{11} A + \ell_{12} (\Theta - T)$$

$$-\dot{S}_V = \ell_{21} + \ell_{22} (\Theta - T)$$

can be readily evaluated from the discussed kinetic model since by definition:

$$-\ell_{11} = \left[\left(\frac{\vartheta \dot{c}}{\vartheta A}\right)_{T,V,\Theta}\right]_{A=0,\ \Theta=T}$$

$$-\ell_{21} = \left[\left(\frac{\vartheta \dot{c}}{\vartheta(\Theta - T)}\right)_{T,V,\Theta}\right]_{A=0,\Theta=T}$$

$$-\ell_{21} = \left[\left(\frac{\vartheta \dot{S}_V}{\vartheta(A)}\right)_{T,V,A}\right]_{A=0,\Theta=T}$$

$$-\ell_{22} = \left[\left(\frac{\vartheta \dot{S}_V}{\vartheta(\Theta - T)}\right)_{T,V,A}\right]_{A=0,\Theta=T}$$

The result is (see Ref. 7) :

$$\ell_{11} = c_f m_2 / R_0 T$$

$$\ell_{12} = \ell_{21} = -\frac{R_0 \ell_{11}}{m_2}\left[\ln Q(T) + \frac{D_0}{2R_0T}\left(1 - \frac{\bar{\Theta} R_0}{D_0}\right)\right]$$

$$\ell_{22} = \frac{1}{tV}\frac{(1-c)}{T}\frac{d}{dT}\left[E(T)\right] + \frac{\ell_{12}^2}{\ell_{11}} +$$

$$+ \frac{\ell_{11} D_0^2}{12 m_2 T^2}\left[1 - \frac{O^2 R^2}{D_0^2}\right]$$

It is seen that the Onsager relations hold and, as it can be checked, that $(\ell_{12}^2 / \ell_{11}\ell_{22}) < 1$.

The complete determination of the explicit forms of the coefficients still required the knowledge of the vibrational characteristic time $t_v$ and of the backward rate "constant" $K_b$. These expressions must be deduced from experiments.

## 14. VECTORIAL FLUXES

### Gas mixtures in thermal equilibrium

We begin considering a mixture of $n$ gasses in thermal equilibrium. This presentation follows closely to one given in (Ref. 3) except for the fact that we do not assume " "mechanical equilibrium" so that $\underline{\nabla} p \neq 0$. Upon eq. (12.1) the contribution to the entropy production due to the vectorial fluxes can be written as:

$$- \dot{S}_1 = \underline{j}_q' \cdot \frac{\underline{\nabla} T}{T^2} + \sum_{i=1}^{n-1} \underline{j}_{c_i} \cdot \underline{F}_i \frac{1}{T} \tag{14.1}$$

with:

$$\underline{F}_i = \underline{\nabla}_T (\varrho_i - \varrho_n) \tag{14.2}$$

The phenomenological equations read:

$$-\underline{j}_p = \frac{L_{pp}}{T^2} \underline{\nabla} T + \sum_{i=1}^{n-1} \frac{L_{pi}}{T} \underline{F}_i$$

(14.3)

$$-\underline{j}_{c_j} = \frac{L_{jp}}{T^2} \underline{\nabla} T + \sum_{i=1}^{n-1} \frac{L_{ji}}{T} \underline{F}_i \qquad (1 \le j \le n-1)$$

and the corresponding Onsager relations are:

$$L_{pi} = L_{ip} \; ; \quad L_{ji} = L_{ij} \qquad (\forall i, j)$$

The diffusion of heat and mass in gas mixtures has been treated very extensively in the pertinent literature. Different (albeit equivalent) sets of phenomenological equations have been studied which are based on different definitions of forces and fluxes (the latter ones, in turn, may be referred, for systems in mechanical equilibrium, to a number of different reference velocities.) It is impossible to give here even a partial account of all this. We shall thus limit ourselves to discuss one particular transformation of the generalized forces which brings forth, explicitly, their dependence upon the mass concentration gradients.

We first use the Gibbs-Duhem equation for $\underline{\nabla} T = 0$

$$-V \underline{\nabla} p + \sum_{i=1}^{n-1} c_i \underline{\nabla}_T g_i = 0$$

to eliminate the quantity $\underline{\nabla}_T g_n$ so that $\underline{F}_i$ becomes:

$$\underline{F}_i = \sum_{j=1}^{n-1} \left( \delta_{ij} + \frac{c_j}{c_n} \right) \underline{\nabla}_T g_j - \frac{V}{c_n} \underline{\nabla} p \tag{14.4}$$

with $\delta_{ij}$ the Kronecker delta. We then consider $g_j$ as a function of the thermodynamic set $\left[T, p, c_i (1 \leq i \leq A-1)\right]$ and express its isothermal gradient as:

$$\nabla_T g_j = \bar{V}_j \underline{\nabla} p + \sum_{l=1}^{n-1} \left( \frac{\partial g_j}{\partial c_e} \right) \underline{\nabla} c_g \tag{14.5}$$

where $\bar{V}_j$ is the partial specific volume defined as:

$$\bar{V}_j \left( \frac{\partial V}{\partial m_j} \right)_{T, p, m_i \neq j} = \left( \frac{\partial g_j}{\partial p} \right)_{T, m_i}$$

The second equality follows from the Maxwell's relation, see eq. (4.3) since the couples $(p, V)$ and $(g, m_i)$ are conjugate with respect to the Gibbs potential $G = V - TS + pV$.

When eq. (14.5) is substituted equation (14.4)

and when one accounts for the fact that, by definition of partial specific quantities:

$$V = \sum_{i=1}^{n} c_i \bar{V}_i$$

one obtains the following expression for the generalized force $\underline{F}_i$ conjugate to $\underline{j}_{ci}$

$$\underline{F}_i = \sum_{j=1}^{n-1} \left( \delta_{ij} + \frac{c_j}{c_n} \right) \sum_{\ell=1}^{n-1} \left( \frac{\partial g_i}{\partial c_\ell} \right) \underline{\nabla}\, c_\ell + (V_i - V_n)\, \underline{\nabla}\, p \qquad (14.6)$$

Thus, as expectable, the "driving force" for the diffusion of mass includes a term proportional to the pressure gradient. [Had we considered gravitational effects the expression for $F_i$ would have included an additional term due to the "body" forces].

If one combines the expression (14.6) for $F_i$ with the phenomenological relations (14.3) one realizes that although there are only $[n(n+1)/2]$ independent phenomenological coefficients one can combine them (both among themselves and with thermodynamic state functions) in a very large number of different ways. The pertinent literature does indeed offer a rather "crowded' picture in this respect. It is thus recommended to pay good attention to the different ways in which phenomenological coefficients are introduced. We shall furnish some such instances for the particular case of a binary mixture. [As discussed in (Ref.8) this approach may be appropriate not only for a truly binary mixture - such as the Lighthill gas - but also for a multi-component mixture in which one can distinguish "light" and "heavy" components and is only interested in the diffusion of any light component with respect to the heavy ones].

For $n = 2$ one diffusion flux independent and the conjugate force $\underline{F}_1$, becomes with : $c_1 + c_2 = 1$

$$\underline{F}_1 = \frac{1}{c_2}\left(\frac{\partial g_1}{\partial c_1}\right)\underline{\nabla} c_1 + (\bar{V}_1 - \bar{V}_2)\underline{\nabla} p \tag{14.7}$$

The phenomenological relations read:

$$-\underline{\dot{J}}_p = \frac{L_{pp}}{T^2}\underline{\nabla} T + \frac{L_{p1} g_{11}}{c_2 T}\left[\underline{\nabla} c_1 + \frac{c_2(\bar{V}_1 - \bar{V}_2)}{g_{11}}\nabla p\right]$$

$$-\underline{\dot{J}}_{c_1} = \frac{L_{p1}}{T^2}\underline{\nabla} T + \frac{L_{11} g_{11}}{c_2 T}\left[\underline{\nabla} c_1 + \frac{c_2(\bar{V}_1 - \bar{V}_2)}{g_{11}}\underline{\nabla} p\right]$$

with:

$$g_{11} = \frac{\partial g_1(T, p, c_1)}{\partial c_1} > 0$$

the inequality following from thermodynamic stability. The positivity of the entropy production implies that:

$$L_{pp} \geq 0 \ ; \quad L_{11} \geq 0 \ ; \quad L_{pp} L_{11} - L_{p1}^2 \geq 0 \tag{14.8}$$

where the last equality sign applies only if $L_{pp} = L_{11} = L_{p1} = 0$

The direct effects of heat conduction and diffusion are measured by the coefficients $L_{pp}$ and $L_{11}$ or, equivalently, by the coefficients:

$$\lambda = L_{pp} / T^2 \qquad \text{(heat conductivity coefficient)}$$

$$D = L_{11} / p c_2 T \qquad \text{(heat coefficient)}$$

The third coefficient measures the cross effects and, when defined as:

$$D' = \frac{L p_1}{p c_1 c_2 T^2}$$

it is usually referred to as the Dufour coefficient. In terms of these new coefficients the phenomenological relations read:

$$-\underline{j}_p = \lambda \underline{\nabla} T + p c_1 g_{11} T D' \left[ \underline{\nabla} c_1 + \frac{c_2 (\bar{V}_1 - \bar{V}_2)}{g_{11}} \underline{\nabla} p \right]$$

$$-\underline{j}_{c_1} = p c_1 c_2 D' \underline{\nabla} T + p D \left[ \underline{\nabla} c_1 + \frac{(\bar{V}_1 - \bar{V}_2)}{g_{11}} \underline{\nabla} p \right]$$

The cross-effect of heat-flow induced by concentration gradients is known as the Dufour effect. The corresponding mass diffusion due to temperature gradients is known as the Soret effect. [Notice that the pressure gradients play the same role as the concentration gradients]. The coefficients $\lambda, D, D'$ are also known as "transport coefficients".

The inequalities (14.8) imply the following inequalities for the transport coefficients:

$$\lambda \geq 0 \; ; \quad D \geq 0 \; ; \quad D'^2 \leq \frac{\lambda D}{T\varrho c_1^2 c_2 g_{11}}$$

since, as seen, $g_{11} > 0$.

Other coefficients often used in the literature are: the "thermal diffusion ratio":

$$K_T = c_1 c_2 \frac{TD'}{D} = L_{p_1} c_2 / L_{11}$$

or the Soret coefficient:

$$S_T = \frac{D'}{D} = \frac{L p_1}{c_1 L_{11} T}$$

Order of magnitude values for these coefficients for gas mixture are (Ref.5)

$$S_T \simeq 10^{-3} \div 10^{-5} (0\,K)^{-1} ; \quad D \simeq 10^{-1}\,\mathrm{cm^2/sec} \; ; \quad D' \simeq 10^{-4}\,\mathrm{cm^2\,sec}(0K)^{-1} ;$$

$$\left[ T \varrho c_2 g_{11} D'^2 / \lambda D \right] \simeq 10^{-2} \div 10^{-3}$$

2. Gas Mixtures in thermal non-equilibrium

Consider now the case of a mixture of $n$ gases in which $(k)$ subsystems are not in thermal equilibrium. Such a case does not seem to have been considered yet in the open literature.

Upon eq. (12.1) the contribution to the entropy production due to the vectorial fluxes is now:

$$-\dot{S}_i = \dot{\underline{j}}'_p \cdot \frac{\underline{\nabla} T}{T^2} + \frac{1}{T} \sum_{i=1}^{n-1} \dot{\underline{j}}_{c_i} \cdot F + \sum_{r=1}^{k} \dot{\underline{j}}_{Sr} \cdot \underline{\nabla} \left( \frac{\Theta_r - T}{T} \right)$$

with the generalized forces $\underline{F}_i$ still given by equations (14.2) The phenomenological equations read:

$$-\dot{\underline{j}}_q = \frac{L_{pp}}{T^2} \underline{\nabla} T + \frac{1}{T} \sum_{i=1}^{n-1} L_{p_i} \underline{F}_i + \sum_{\beta=1}^{k} L_{p,S\beta} \underline{\nabla} \left( \frac{\Theta_\beta - T}{T} \right)$$

$$-\dot{\underline{j}}_{cr} = \frac{L_{rp}}{T^2} \underline{\nabla} T + \frac{1}{T} \sum_{i=1}^{n-1} L_{r_i} \underline{F}_i + \sum_{\beta=1}^{k} L_{c_r,S\beta} \underline{\nabla} \left( \frac{\Theta_\beta - T}{T} \right)$$

$$(1 \leq r \leq n-1) \qquad (14.9)$$

$$-\underline{j}_{s\alpha} = \frac{L_{s\alpha,q}}{T^2}\,\underline{\nabla}\,T + \frac{1}{T}\sum_{i=1}^{n-1} L_{s,\alpha,i}\,\underline{F}_i + \sum_{\beta=1}^{k} L_{s\alpha,s\beta}\,\underline{\nabla}\left(\frac{\Theta_p - T}{T}\right)$$

$$(1 \leq \alpha \leq K)$$

with

$$L_{rp} = L_{pr} \qquad (1 \leq \forall r \leq n-1)$$

$$L_{ri} = L_{ir} \qquad (1 \leq \forall r, i \leq n-1)$$

$$L_{p,s\alpha} = L_{s\alpha,p} \qquad (1 \leq \forall\, \alpha \leq K)$$

$$L_{cr,s\beta} = L_{s\beta,cr} \qquad (1 \leq \forall r \leq n-1;\ 1 \leq \forall \beta \leq K)$$

$$L_{s\alpha,s\beta} = L_{s\beta,s\alpha} \qquad (1 \leq \forall \alpha, \beta \leq K)$$

upon the Onsager relation. As always, the postulate of positive entropy production implies the positive definite character of the matrix of the phenomenological coefficients.

As in the case of thermal equilibrium, we now transform the generalized forces. The Gibbs-Duhem equation for $\underline{\nabla}\,T = 0$ reads:

$$-V\,\underline{\nabla}\,p + \sum_{i=1}^{n-1} c_i\,\underline{\nabla}_T\,\varrho_i + \sum_{r=1}^{k} S_r\,\underline{\nabla}(\Theta_r - T) = 0$$

that elimination of $\underline{\nabla}_T \varrho_n$ from equation (14.2) leads to:

$$\underline{F}_i = \sum_{j=1}^{n-1} \left( \vartheta_{ij} + \frac{c_j}{c_n} \right) \underline{\nabla}_T \varrho_j - \frac{V}{c_n} \underline{\nabla} p + \sum_{r=1}^{K} \frac{S_r}{c_n} \underline{\nabla} (\Theta_r - T) \tag{14.10}$$

We have now to consider $\varrho_i$ as function of the thermodynamic set

$$T, p, c_i \, (1 \leq i \leq n-1), \, (\Theta_r - T), \, (1 \leq r \leq K)$$

so that its isothermic gradient can be expressed as:

$$\underline{\nabla}_T \varrho_j = \hat{V}_j \, \underline{\nabla} p - \sum_{r=1}^{K} (\hat{S}_r)_j \, \underline{\nabla}(\Theta_r - T) + \sum_{\ell=1}^{n-1} \left( \frac{\vartheta \varrho_i}{\vartheta c_\ell} \right) \underline{\nabla} c_\ell \tag{14.11}$$

where $\hat{V}_j$ and $(\hat{S}_r)_j$ are the (generalized) partial specific volume and entropy $S_r$ [see eq. (5.9)]. The thermodynamic relation (14.11) follows from the equalities:

$$\left( \frac{\vartheta V}{\vartheta M_j} \right)_{T, p, \Theta_r, M_{i \neq j}} = \left( \frac{\vartheta \varrho_i}{\vartheta p} \right)_{T, \Theta_r, M_i}$$

$$- \left( \frac{\vartheta S_r}{\vartheta M_j} \right)_{T, p, \Theta, M_{i \neq j}} = \left[ \frac{\vartheta \varrho_i}{\vartheta (\Theta_r - T)} \right]_{T, p, \Theta_{s \neq r} - T, M_i}$$

which represent sets of Maxwell's relations [see eqs. (4.3)] since the couples $(p, V), (\varrho_i, M_j), [-S_r (\Theta_r - T)]$ are conjugate with respect to the generalized Gibbs potential.

$$\underset{\sim}{G} = u - TS - pV - \sum_{r=1}^{k} s_r(\Theta_r - T) = \sum_{i=1}^{n} \varrho_i M_i$$

Upon the definition of (generalized) partial specific quantity it is see eq. (5.10)

$$q = \sum_{n=1}^{n} c_i \hat{Q}_i$$

It then follows that:

$$\sum_{j=1}^{n-1} \left( \delta_{ij} + \frac{c_j}{c_n} \right) \hat{Q}_j = \hat{Q}_i + \frac{q}{c_n} - \hat{Q}_n$$

Hence, substitution of eq. (14.11) into eq. (14.10) yields:

$$\underline{F}_i = \sum_{j=1}^{n-1} \left( \delta_{ij} + \frac{c_j}{c_n} \right) \sum_{\ell=1}^{n-1} \left( \frac{\vartheta \varrho_i}{\vartheta c_\ell} \right) \underline{\nabla} c_\ell +$$

$$+ \left( \hat{V}_i - \hat{V}_n \right) \underline{\nabla} p + \sum_{r=1}^{k} \left[ (\hat{S}_r)_n - (\hat{S}_r)_i \right] \underline{\nabla} (\Theta_r - T) \tag{14.12}$$

Comparison between eqs. (14.12) and (14.6) shows that, in conditions of thermal non-equilibrium the generalized force conjugate to the mass diffusion fluxes exhibits the following differences, as compared to the case of thermal equilibrium :

i) the generalized partial specific quantities appear

ii) in addition to contributions due to pressure gradients there are those due to the gradients of the temperature differences $(\Theta_r - \Theta)$.

To simplify further discussions, consider the case of the ideal biatomic dissociating and vibrating gas. In this case we have one dependent mass diffusion flux and one flux representing the "diffusion" of energy between the subsystems in thermal non-equilibrium.

From the state equations (6.21) one obtains:

$$\hat{v}_1 - \hat{v}_2 = \frac{R_0 T}{m_2 p}$$

$$(\hat{s}_1)_2 = s_v = \frac{R_0}{m_2} \frac{(1-c)}{2} \left[1 + \ln(\Theta / \bar{\Theta})\right]$$

$$(\hat{s}_1) = 0$$

so that (with $c_1 = c$; $\Theta_1 = \Theta$ ):

$$\underline{F}_1 = \frac{1}{1-c} g_{11} \underline{\nabla} c + \frac{R_0 T}{m_2 p} \underline{\nabla} p + s_v \underline{\nabla} (\Theta - T)$$

and the phenomenological relations become, on account of the Onsager relations:

$$-\underline{j}'_{qq} = \frac{L_{qq}}{T^2} \underline{\nabla} T + \frac{1}{T} L_{q1} \underline{F}_1 + L_{q,sv} \underline{\nabla} \left(\frac{\Theta - T}{T}\right)$$

$$-\underline{j}_c = L_{q1}\frac{\underline{\nabla} T}{T^2} + \frac{L_{11}}{T}\underline{F}_1 + L_{1,sv}\underline{\nabla}\left(\frac{\Theta - T}{T}\right)$$

$$-\underline{j}_{sv} = L_{q,sv}\frac{\underline{\nabla} T}{T^2} + \frac{L_{1,sv}}{T}\underline{F}_1 + L_{sv,sv}\underline{\nabla}\left(\frac{\Theta - T}{T}\right)$$

with:

$$L_{qq} \geq 0 \qquad L_{11} \geq 0 \qquad L_{sv,sv} \geq 0$$

$$L_{qq}L_{11} - L_{q1}^2 > 0 \qquad L_{qq}L_{sv,sv} - L_{q,sv}^2 > 0$$

$$L_{11}L_{sv,sv} - L_{1sv}^2 > 0$$

$$\det\begin{bmatrix} L_{qq} & L_{q1} & L_{q,sv} \\ L_{q1} & L_{11} & L_{1,sv} \\ L_{q,sv} & L_{1,sv} & L_{sv,sv} \end{bmatrix} > 0$$

Besides direct diffusion of heat and mass, and besides the Dufour and Soret cross-effects we see that there are a number of other cross effects and a direct diffusion of "heat" between the subsystems in thermal non-equilibrium. Hardly anything is known, theoretically or experimentally, about these additional cross-effects and their orders of magnitude. This is a field completely open for further challenging researches.

## 15. PARTICULAR CASE. SIMPLE GAS.

As seen, in the case of a simple gas the induced unknowns reduce to:

1) the scalar quantity

$$(\pi - p)$$

giving the excess average normal stress over the thermodynamic pressure ;

2) The tensorial quantity:

$$\underline{\underline{\tau}}_0$$

which is traceless and symmetric;

3) The vectorial quantity:

$$\underline{j}_q = T\underline{j}_s = \underline{j}_n$$

where $\underline{j}_q$ is the heat flux.

To within a linear phenomenology one lets:

$$(\pi - p) = -\mu_2 \underline{\nabla} \cdot \underline{V}$$

$$\underline{\underline{\tau}}_0 = -2\mu(\underline{\nabla}\,\underline{V})_0 \tag{15.1}$$

$$\underline{j}_q = -\lambda \underline{\nabla} T$$

where $(\underline{\nabla}\,\underline{V})_0$ is the traceless symmetry part of the dyadic $\underline{\nabla}\,\underline{V}$ Eq. $(15.1)_2$ and $(15.1)_3$ are, respectively, the Newton and Fourier laws.

In equations (15.1) the "kinetic" (or transport)

coefficients are state functions (for gases in usual pressure and temperature ranges they are practically functions only of the absolute temperature $T$). Specifically: $\mu$ is the coefficient of viscosity; $\mu_2$ is the <u>second</u> coefficient of viscosity and $\lambda$ is the coefficient of thermal conduction.

For the entropy production one has:

$$T\dot{s} = -(\pi - p)\,\underline{\nabla}\,\underline{V} - \underline{j}_q \cdot \frac{\underline{\nabla} T}{T} - \underline{\underline{\tau}}_0 : (\underline{\nabla}\,\underline{V})_0 =$$

(15.2) $$= \mu_2 (\underline{\nabla}\,\underline{V})^2 + \frac{\lambda}{T} (\underline{\nabla} T)^2 + 2\mu (\underline{\nabla}\,\underline{V})_0 : (\underline{\nabla}\,\underline{V})_0$$

From the postulate of positive entropy production it follows that:

(15.3) $$\mu \geq 0 \quad ; \quad \mu_2 \geq 0 \quad ; \quad \lambda \geq 0$$

Hence, besides the well known statement "heat flows spontaneously from the regions of higher temperature to those of lower temperature" one can formulate a similar statement for the flux of momentum. Namely:

"As a consequence of the positivity of entropy production momentum flows spontaneously from the regions of higher velocities to those of lower velocities".

In terms of cartesian components, if $(u, v, w)$ are the velocity components, $\sigma_x, \sigma_y, \sigma_z$ the (total) normal stresses and $\tau_{xy}, \tau_{yz}, \tau_{xy}$ the tangential stresses equations (15.1) and $(15.1)_2$ become:

$$\pi = p + \mu_2 (u_x + v_y + w_z) = \frac{\sigma_x + \sigma_y + \sigma_z}{3}$$

$$\sigma_x - p = -2\mu \left[ u_x - \frac{1}{3}(u_x + v_y + w_z) \right]$$

$$\tau_{xy} = -\mu (u_y + v_x)$$

with similar expressions for the other stress components. Combining the first two equations yields:

$$\sigma_x = p - 2\mu u_x + \left(\mu_2 + \frac{2}{3}\mu\right) \underline{\nabla} \cdot \underline{V}$$

a relation which shows close formal analogy with Hooke's law (notice, however, that the latter is an equation of state) with $\mu$ and $\frac{2}{3}\mu + \mu_2$ playing the role of the Lame's constants.

For usual gases it is $\frac{3\mu_2}{2\mu} \ll 1$ and the contribution of the second coefficient of viscosity can be usually neglected. In these cases the average normal stresses is taken to be equal to the thermodynamic pressure $p$ (i.e. one takes $\mu_2 = 0$).

## 16. SUMMARY OF THE CLOSED SET OF BASIC FIELD EQUATIONS AND OTHER RELEVANT EQUATIONS

In a gas mixture of $n$ gases wherin $k$ ensembles of internal defrees of freedom are not in thermal equilibrium and a number $r$ of independent reactions takes place the fluid-dynamic field unknowns, in a single fluid theory, are the mass velocity $V$ of the mixture plus a set of $d = (n + K - 1)$ independent thermodynamic specific and/or intensive quantities. If one takes for this set the natural variables for the internal energy, the primitive field unknowns are:

$$s, \; v = \frac{1}{\rho}, \quad c_i \; (1 \le i \le n-1), \quad s_j \; (1 \le j \le k), \; \underline{V}$$

The closed set of equations for the determination of these unknowns is given by:

1) A thermodynamic fundamental relation or, equivalently, $d$ independent equations of state. With the energy representation of the system the fundamental relation is:

$$u = u(s, v, c_i, s_j)$$

where $u$ is the specific internal energy.

A set of equation of state is given by the derivatives of $u$, according to the Gibbs relation:

$$du = T ds - p dV + \sum_{i=1}^{n-1} (g_i - g_n) dc_i + \sum_{j=1}^{n} (\theta_j - T) ds_j$$

which defines the set of intensive parameters:

$$T; \; -p; \; (g_i - g_n) \; (1 \le i \le n-1); \; (\theta_j - T) \; (1 \le j \le k)$$

conjugated to the parameters: $[s, v, c_i, s_j]$.

2) The balance and/or conservation equations:

$$\varrho \frac{ds}{dt} + \underline{\nabla} \cdot \underline{J}_s = \dot{s} \qquad \left[\frac{d}{dt} = \frac{\partial}{\partial t} + \underline{V} \cdot \underline{\nabla}\right]$$

$$\varrho \frac{dv}{dt} = \underline{\nabla} \cdot \underline{V}$$

$$\varrho \frac{dc_i}{dt} + \underline{\nabla} \cdot \underline{J}_{c_i} = \dot{c}_i \qquad (1 \leq i \leq n-1)$$

$$\varrho \frac{ds_j}{dt} + \underline{\nabla} \cdot \underline{J}_{s_j} = \dot{s}_j \qquad (1 \leq S \leq k)$$

$$\varrho \frac{d\underline{V}}{dt} + \underline{\nabla} \cdot \left[p\,\underline{\underline{U}} + (\pi - p)\,\underline{\underline{U}} + \underline{\underline{\tau}}_0\right] = 0 \qquad (\tilde{\underline{\underline{\tau}}}_0 = \underline{\underline{\tau}}_0)$$

3) The expressions for the mass productions $\dot{c}_i$ in terms of the rates of change $\dot{\xi}_\ell$ of the r-th independent progress variables. If the r independent chemical reactions are symbolically indicated as:

$$\sum_{i=1}^{n} \bar{\nu}_{\ell i}\left[x_i\right] = \sum_{i=1}^{n} \bar{\nu}'_{\ell i}\left[x_i\right]$$

it is:

$$\dot{c}_i = \sum_{\ell=1}^{r} \nu_{\ell i}\, \dot{\xi}_\ell \;; \qquad \left[\sum_{i=1}^{n} \nu_{\ell i} = 0, \;\forall \ell\right]$$

where

$$\nu_{\ell i} = \frac{(\bar{\nu}'_{\ell i} - \nu_{\ell i})\, m_i}{m_r}$$

is the molar mass of the $i$-th constituent, $m_r$ a reference molar mass, $\bar{\nu}_{\ell i}$ (resp. $\bar{\nu}'_{\ell i}$ ) the stoichiometric coefficients of the reactants (resp. reaction products) in the $\ell$-th reaction.

4) The expression for the entropy production:

$$-T\dot{s} = (\pi - p)\,\underline{\nabla}\cdot\underline{V} + \sum_{\ell=1}^{r} A_\ell\, \dot{\xi}_\ell + \sum_{j=1}^{k} (\Theta_j - T)\dot{s}_j +$$

$$+ \underline{J}'_q \cdot \frac{\underline{\nabla} T}{T} + \sum_{i=1}^{n-1} \underline{J}_{c_i}\cdot \underline{\nabla}_T (g_i - g_n) + \sum_{j=1}^{k} \underline{J}_s \cdot T\, \underline{\nabla}\left(\frac{\Theta_j - T}{T}\right) +$$

$$+ \underline{\underline{\tau}}_0 : (\underline{\nabla}\, \underline{V})_0 = -T(\dot{s}_0 + \dot{s}_1 + \dot{s}_r) \leq 0$$

where $\dot{s}_i$ is the entropy production corresponding to generalized fluxes of the $i$-th tensorial order and $\underline{\nabla}_T$ the gradient at $T =$ const.

The affinity $A_\ell$ of the $\ell$-th reaction is defined by:

$$A_\ell = \sum_{i=1}^{n} \nu_{\ell i}\, g_i \qquad (1 \leq \ell \leq r)$$

The heat flux $\underline{j}'_q$ is defined by:

$$\underline{j}_q = T\,\underline{J}_s - \sum_{i=1}^{n-1} T\left(\hat{s}_i - \hat{s}_n\right)\underline{J}_{c_i} + \sum_{j=1}^{k}\left(\theta_j - T\right)\underline{j}_{s_j}$$

with $\hat{s}_i$ the generalized partial specific entropy:

$$\hat{s}_i = \left(\frac{\partial s}{\partial m_i}\right)_{T,p,\theta_j,m_{t\neq i}} \qquad \begin{matrix}(1 \leq j \leq K)\\(1 \leq t \leq n-1)\end{matrix}$$

$m_i$ being the mass of the $i$-th component. Equivalently, the heat flux $J'_q$ can be defined in terms of the diffusive fluxes of internal energy $\underline{J}_u$ and of mass by:

$$\underline{J}'_q = \underline{J}_u - \sum_{i=1}^{n} \hat{\underset{\sim}{h}}_1\,\underline{J}_{c_i}$$

where

$$\hat{\underset{\sim}{h}}_i = \left[\frac{\partial\left(m\,\underset{\sim}{h}\right)}{\partial M_i}\right]_{T,p,\theta_j,m_{t\neq i}}$$

$M = \sum_{i=1}^{n} m_i$ is the total mass of the mixture and $\underset{\sim}{h}$ the generalized enthalpy, per unit mass:

$$\underset{\sim}{h} = u + pv - \sum_{j=1}^{k} s_j\left(\theta_j - T\right) = Ts + \sum_{i=1}^{n-1}\left(g_i - g_n\right)c_i =$$

$$= \underset{\sim}{h}\left(s, p, c_i, \theta_j - T\right)$$

5) The expression for the rates $\dot{\xi}_l$ and $\dot{s}_j$ for the difference

between the mean normal stress and the thermodynamic pressure, for the diffusive fluxes $\underline{j}'_q, \underline{j}_{c_i}, \underline{j}_{s_j}$ and for the symmetric, traceless, stress tensor $\underline{\underline{\tau}}_0$

In general:

$$\dot{\xi}_\ell = f \quad \text{(thermodynamic state)}$$

$$\dot{s}_\ell = f \quad \text{(thermodynamic state)}$$

To within a linear phenomenology and for isotropic media:

$$\underline{\underline{\tau}}_0 = -2\mu\left(\underline{\nabla}\,\underline{V}\right)_0$$

$$\begin{bmatrix} \dot{\xi}_1 \\ \vdots \\ \dot{\xi}_r \\ \vdots \\ \dot{s}_1 \\ \vdots \\ \dot{s}_k \\ \pi - p \end{bmatrix} = \left[\ell_{ij}\right] \begin{bmatrix} A_1 \\ \vdots \\ A_r \\ (\Theta_1 - T) \\ \vdots \\ (\Theta_k - T) \\ \underline{\nabla}\cdot\underline{V} \end{bmatrix} ; \; (1 \le i,j \le k+r+1) ; \; \begin{bmatrix} \underline{j}'_q \\ \underline{j}_{c_1} \\ \vdots \\ \underline{j}_{c_{n-1}} \\ \underline{j}_{s_1} \\ \vdots \\ \underline{j}_{s_k} \end{bmatrix} = \left[L_{ij}\right] \begin{bmatrix} \underline{\nabla} T / T^2 \\ \frac{i}{T}\underline{\nabla} T(g_1 - g_n) \\ \vdots \\ \frac{1}{T}\underline{\nabla} T(g_{n-1} - g_n) \\ \underline{\nabla}\left[(\Theta_1 - T)/T\right] \\ \vdots \\ \underline{\nabla}\left[(\Theta_k - T)/T\right] \end{bmatrix}$$

The phenomenological matrices $\underline{\underline{\ell}} \equiv \left[\ell_{ij}\right]$ of order $(k+r+1)$, and $\underline{\underline{L}} \equiv \left[L_{ij}\right]$ are positive definite (postulate of positive entropy production) and symmetric (Onsager postulate):

$$\ell_{ij} = \ell_{ji} \; ; \; L_{ij} = L_{ji} \; ; \qquad (\forall\, i,j)$$

6) The explicit functional dependence of the phenomenological

coefficients $\mu$, $\ell_{ij}$ and $L_{ij}$ (and of the rates $\dot{\xi}_\ell$ and $\dot{s}_j$ if a non-linear phenomenology is to be used for them) upon the thermodynamic state parameters of the mixture.

We shall now list a number of additional equations and/or relations which are either relevant per se or can be used to replace an equivalent number of equations in the above closed set of equations.

7) Energy conservation equation (which can be used to replace the entropy balance equations):

$$\varrho \frac{d}{dt}\left[h + \frac{V^2}{2}\right] - \frac{\partial p}{\partial t} + \underline{\nabla} \cdot \left[\underline{J}_q' + \sum_{i=1}^{n} \underset{\sim}{\hat{h}}_i \underline{J}_{c_i} + \right.$$

$$\left. + (\pi - p)\underline{V} + \underline{\underline{\tau}}_0 \cdot \underline{V}\right] = 0$$

where:

$$h = u + pv = h(s, p, c_i, s_j)$$

is the specific enthalpy:

8) Balance of kinetic energy:

$$\varrho \frac{dV^2/2}{dt} + \underline{\nabla} \cdot \left[p\underline{V} + (\pi - p)\underline{V} + \underline{\underline{\tau}}_0 \cdot \underline{V}\right] =$$

$$= p\underline{\nabla} \cdot \underline{V} + \left[(\pi - p)\underline{\nabla} \cdot \underline{V} + \underline{\underline{\tau}}_0 : (\underline{\nabla}\,\underline{V})_0\right]$$

9) Alternate possible definitions of heat flux:

$$\underline{J}_q = \underline{J}_u = T\underline{J}_s + \sum_{i=1}^{n-1} (g_i - g_n)\underline{J}_{c_i} + \sum_{j=1}^{k} (\theta_j - T)\underline{J}_{s_j}$$

When this definition is used, the appropriate corresponding form of the energy equation is:

$$\varrho \frac{d}{dt}\left(h + \frac{V^2}{2}\right) - \frac{\partial p}{\partial t} + \underline{\nabla}\cdot\left[\underline{J}_q + (\pi - p)\underline{V} + \underline{\underline{\tau}}_0\cdot\underline{V}\right] = 0$$

and in the expression for the entropy production $\dot{s}$ the contribution $\dot{s}_1$ due to vectorial generalized fluxes is to be replaced by:

$$-T\dot{s}_1 = \underline{J}_q\cdot\frac{\underline{\nabla} T}{T} + \sum_{i=1}^{n-1} \underline{J}_{c_i}\cdot\underline{\nabla}\left(\frac{g_i - g_n}{T}\right) + \sum_{j=1}^{k} \underline{j}_{s_j}\cdot T\underline{\nabla}\left(\frac{\theta_j - T}{T}\right)$$

10) The $(n-1)$ mass balance equations can be substituted by $r$ balance equations for the progress variables $\xi_\ell$ and $m-1=n-r-1$ conservation equations for the atomic species present in the mixture:

$$\varrho \frac{d\xi_\ell}{dt} + \underline{\nabla}\cdot j_{\xi_\ell} = \dot{\xi}_\ell \qquad (1 \le \ell \le r)$$

$$\varrho \frac{d\sigma_j}{dt} + \underline{\nabla}\cdot \underline{j}_{\sigma_j} = 0 \quad (1 \le j \le n-r-1 = m-1)$$

where:

$$\xi_l = \sum_{i=1}^{n} M_{li}\left[c_i - c_{io}\right]$$

$$\sigma_j = \sum_{i=1}^{n} M_{ji}\left[c_i - c_{io}\right]$$

the quantities $c_{io}$ are arbitrary constant values of the mass concentration and the non singular squared matrix $\underline{\underline{M}}$ of order $n$ is defined by:

$$\left[M_{ij}\right] = \underline{\underline{M}} = \begin{bmatrix} \tilde{\underline{\underline{R}}}^{-1} & \underline{\underline{0}} \\ -\tilde{\underline{\underline{Q}}}\cdot\tilde{\underline{\underline{R}}}^{-1} & \underline{\underline{U}} \end{bmatrix}$$

Here $\underline{\underline{0}}$ is the $(r \times r)$ null matrix, $\underline{\underline{U}}$ the $(m \times m)$ unit matrix, $\underline{\underline{R}}$ is a non singular square matrix of order $(r)$ and $\underline{\underline{Q}}$ a matrix of order $(r \times m)$. They are submatrices defined as:

$$\underline{\underline{N}} = \left[\underline{\underline{R}} \quad \underline{\underline{Q}}\right]$$

of the matrix $N_{li} = \nu_{li}\left[1 \le l \le r;\ 1 \le i \le n\right]$ of the reduced stoichiometric coefficients $\nu_{li}$.

The fluxes $\underline{j}_{\xi_l}$ and $\underline{j}_{\sigma_j}$ are related to the mass diffusion fluxes by:

$$\underline{J}_{\xi_l} = \sum_{s=1}^{r} \left(\tilde{\underline{\underline{R}}}^{-1}\right)_{ls} \underline{J}_{cs} \qquad (1 \le l \le r)$$

$$\underline{J}_{\sigma j} = \underline{J}_{c_{r+j}} - \sum_{s=1}^{r} \left( \underline{\underline{\tilde{Q}}} \cdot \underline{\underline{\tilde{R}}}^{-1} \right)_{js} \underline{J}_{cs} \qquad (1 \leq j \leq m-1)$$

When these fluxes are employed, the term $\sum_{i=1}^{n-1} \underline{J}_{ci} \cdot \underline{\nabla}_T (g_i - g_n)$ in the expression for the entropy production must be replaced by the terms:

$$\sum_{\ell=1}^{r} \underline{J}_{\xi_\ell} \cdot \underline{\nabla}_T A_\ell + \sum_{j=1}^{m-1} \underline{J}_{\sigma_j} \cdot \underline{\nabla}_T (g_j - g_n)$$

The Onsager relations are still applicable. A corresponding change must be made in the energy conservation equation.

16.1 Particular case. Simple gas.

For a simple gas one has:

- Primitive field unknowns:

$$s, \; v = \frac{1}{\varrho}, \; V$$

- Fundamental relation. Gibbs relation:

$$u = u(s, v)$$

$$du = T ds - p dv$$

- Balance and/or conservation equations:

$$\varrho \frac{D v}{D t} = \underline{\nabla} \cdot \underline{V}$$

$$\varrho \frac{D \underline{V}}{D t} + \underline{\nabla} \cdot \left\{ \left( p - \mu_2 \underline{\nabla} \cdot \underline{V} \right) \underline{\underline{U}} - 2\mu \left( \underline{\nabla}\, \underline{V} \right)_0 \right\} = \varrho \underline{g}$$

$$\varrho \frac{D s}{D t} - \underline{\nabla} \cdot \left[ \frac{\lambda \underline{\nabla} T}{T} \right] = \frac{1}{T} \left\{ \mu_2 \left( \underline{\nabla} \cdot \underline{V} \right)^2 + \right. \qquad (16.1)$$

$$\left. + \frac{\lambda}{T} \left( \underline{\nabla} T \right)^2 + 2\mu \left( \underline{\nabla}\, V \right)_0 : \left( \underline{\nabla}\, \underline{V} \right)_0 \right\}$$

- Conservation of energy:

$$\varrho \frac{D}{D t} \left[ h + \frac{V^2}{2} + \Psi \right] - \frac{\partial p}{\partial t} + \underline{\nabla} \cdot \left[ -\lambda \underline{\nabla} T - \mu_2 \left( \underline{\nabla} \cdot \underline{V} \right) \underline{V} + \right.$$

$$\left. - 2\mu \left( \underline{\nabla}\, \underline{V} \right)_0 \cdot \underline{V} \right] = 0 \qquad (16.2)$$

The systems of equations (16.1) is a closed one, once the functions $h = h(s, v)$; $u = u(s, v)$; $\mu_2 = \mu_2(s, v)$; $\mu = \mu(s, v)$ have been assigned.

For a perfect gas with constant specific heats one may close the system of equations by prescribing, equivalently, the two equations of state.

$$p v = R T \; ; \qquad u = c_v T \qquad (c_v = \text{const})$$

For most practical purposes, as said, one may take $\mu_2 = 0$ and $\mu$ and $\lambda$ functions only of $T$. Even more simply, whenever the temperature variations in the flow field are not large, one may take $\mu$ and $\lambda$ constants.

For continuous motion the energy equation (16.2) is not independent from the other three equations (16.1) and as basic system one may take the continuity, momentum and energy equations.

## 17. ANALYSIS OF DISCONTINUITIES

Let us consider, for simplicity, the case of a simple gas.

The pertinent set of jump equations is then:

Mass $$\delta\left[\underline{u}\cdot(\varrho V)\right] = 0$$

(17.1)

Total energy $$\delta\left[\underline{u}\cdot\left\{\varrho V H + \underline{\tau}\cdot\underline{V} + \zeta_q\right\}\right] = 0$$

Momentum $$\delta\left[\underline{u}\cdot\left(\varrho \underline{V}\,\underline{V} + p\,\underline{\underline{U}} + \tau_0\right)\right] = 0$$

For continuous motions the entropy balance equation is not independent of those of mass, momentum and energy. This is not necessarily so, for discontinuous motion and the entropy jump equation.

$$\delta\left[u\cdot\left(\varrho\underline{V}s + \underline{j}_s\right)\right] = \dot{s}_s > 0$$

may constitute an additional independent statement.

The jump conditions were derived upon the assumption that the total fluxes (and, hence, the diffusive fluxes) remained finite. Even to within a linear phenomenology, the diffusive fluxes depend upon gradients of the basic field variables. Hence, if they have to remain finite a necessary (albeit not sufficient) condition is that the field variables be everywhere continuous.

Thus, strictly speaking, no discontinuity may be present in a real fluid. It may occur, however, that the basic field variables undergo very large changes in space regions in which one dimension is very small compared to the other two. In these cases it may prove convenient, to neglect altogether the "thickness" of these regions and treat the regions themselves as discontinuity surface.

The analysis therefore, is to be carried out in two subsequent steps. One first neglects by assumption, the terms which would certainly prevent the existence of discontinuity surfaces (namely the diffusive, irreversible, fluxes) investigates whether any discontinuity surface can exist in such conditions (which obviously, are those pertaining to an ideal fluid) and analyses their properties. Subsequently, one determines whether (and to what extent) such discontinuities can still be used (as a mathematical abstraction) in a real fluid. It turns out, as we shall presently see, that there are two different classes of discontinuities: the shock discontinuities and the contact and/or vortex discontinuities. The first ones can be dealt with as such, even in non-ideal fluids, as they correspond to physical realities. As a matter of fact, they do exist only if the fluid is non ideal. In other words, it does happen that the rapid changes in field variables are "permanently" limited within a very thin region whose "thickness" is of the order of the free molecular path. Hence, at least in the continuum regime and provided one is not interested in the details of the flow field within the shock, the region itself can be treated as a discontinuity surface.

The matter is different for the second type of discontinuities. In a real fluid they are "resolved" into regions (so called dissipative regions) which can no longer be approximated as discontinuity surfaces. Yet in many cases (i.e. whenever the ratio between diffusive and convective fluxes is very small in the rest of the flow field) one can still take the results of the ideal flüid theory (and, hence, of the discontinuities) as a relevant first approximation for the flow field and analyse, in a subsequent stage (if needed) the details of the dissipative regions. (When this procedure is applicable one says that the "boundary layer" approximation is valid).

To proceed, then, to the first stage of the analysis, we write the jump equations (17.1) for an ideal fluid as:

$$\delta\left[\underline{u}\cdot(\varrho\,\underline{V})\right] = 0$$

$$\delta\left[\underline{u}\cdot(\varrho\underline{V}H)\right]=0$$

$$\delta\left[\underline{u}\cdot(\varrho VV+p\underline{\underline{U}})\right]=0 \tag{17.2}$$

$$\delta\left[\underline{u}\cdot(\varrho\underline{V}s)\right]=\dot{s}_s$$

The first one implies that the mass flux $(\varrho V_n)$ through any discontinuity surfaces is necessarily continuous. Let $m=\varrho V_n$ be this mass flux and let $\underline{t}$ be any unit vector lying on the discontinuity surface $\sigma$ (thus $\underline{n}\cdot\underline{t}=0$ by position). By multiplying the momentum jump condition scalarly by $\underline{n}$ and $\underline{t}$ one will then obtain: $\left[\text{since } \delta(fg)=f\delta g \text{ when } \delta f=0\right]$:

$$\delta_m=0$$

$$m\,\delta H=0$$

$$m\,\delta V_n+\delta p=0 \qquad \left(H=h+\frac{V^2}{2}=h+\frac{V_n^2}{2}+\frac{V_t^2}{2}\right)$$

$$m\,\delta\underline{V}_t=0 \tag{17.3}$$

$$m\,\delta s=\dot{s}_s\geq 0$$

where $\underline{V}_t$ is the component of $\underline{V}$ in the direction of $\underline{t}$. Two seperate cases need to be investigated.

I) $m=\varrho V_n=0$: no mass flux crosses the discontinuity surface which, therefore (since $\varrho\neq 0$), is a stream surface (i.e. the velocity vector $\underline{V}$ is tangent to the surface on both sides).

In this case eqs. $(17.2)_{2,4}$ are identically satisfied for any $\delta H$ and $\delta \underline{V}_t$ whereas equation $(17.2)_{3,5}$ yields:

$$\delta p = 0$$

$$\dot{s}_s = 0$$

the last one holding for any $\delta s$.

- The surface entropy production is necessarily zero.

  Whenever one (or more) of the quantities $\delta H, \delta \underline{V}_t, \delta s$ is different from zero, the surface $\sigma$ is a discontinuity surface.

  For these types of discontinuities:

- The pressure is always continuous
- The surface entropy production is always zero
- There may exist discontinuities in the tangential velocity $\underline{V}_t$ and/or in thermodynamic quantities other than the pressure.
- Upon expliciting the expression for $H$ one gets:

$$\delta H = \delta u + p\,\delta\left(\frac{1}{\varrho}\right) + \delta\left(\frac{V_t^2}{2}\right)$$

- The jump of only one thermodynamic quantity is independent, the expression for the other ones follows from the continuity of $p$ and the equations of state [thus, for instance, for a perfect gas it is $\delta T = \frac{1}{c_p}\delta h = \frac{p}{R}\delta\left(\frac{1}{\varrho}\right) = \frac{\delta u}{c_v}$ ].

- When the tangential velocity is continuous $(\delta \underline{V}_t = 0)$ the discontinuity surface is called "contact surface" since it separates two different thermodynamic isobaric states of the medium. [More generally, contact surfaces may separate two "different" fluids. Recall that contact surfaces are necessarily stream surfaces].
- When the tangential velocity is discontinuous, the surface $\sigma$ is called a "vortex surface". [The thermodynamic states on the two sides of a vortex surface may or may not be different]. This terminology follows from the fact that when $\delta \underline{V}_t \neq 0$ the vector $\underline{\omega} = \underline{\nabla} \wedge \underline{V}$ is tangent to the surface $\sigma$ and assumes, on it, an infinitely large value. The proof of these statements hinges on the Stokes theorem. To begin with there can be no component of $\underline{\omega}$ normal to $\sigma$ since, on $\sigma$, $V_n = 0$ and, therefore, the circulation around any closed contour lying on $\sigma$ is zero. Consider now (see figure) the intersection between the surface $\sigma$ and a plane normal to the unit vector $\underline{t}$ in P. Application of the Stokes theorem to the circuit shown in the figure leads to:

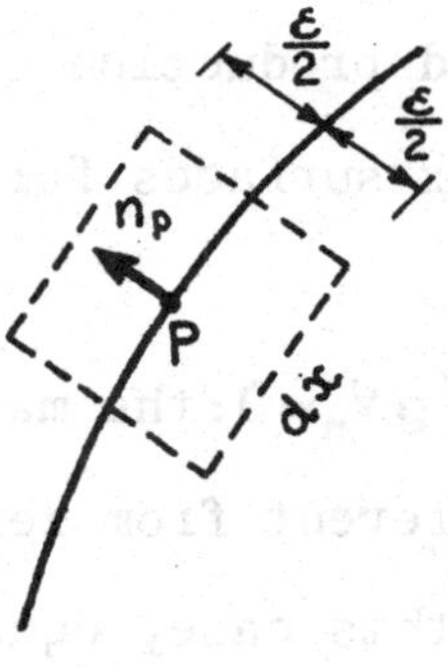


$$\lim_{\varepsilon \to 0} \int_{-\frac{\varepsilon}{2}}^{\frac{\varepsilon}{2}} \underline{t} \cdot \underline{\omega} \, dn = \delta \underline{V}_t$$

which proves that the surface $\sigma$ is a surface of infinite "vorticity" .

- The discontinuity of the tangential component of the velocity across $\sigma$ can be in direction and/or in intensity.
- Contact and/or vortex surfaces are mathematical abstractions for "ideal" fluids. Their use in connection with real fluids as a first approximation for the flow field variables is admissible when the overall flow conditions are such that the ratio between the diffusive and convective fluxes is negligible almost everywhere. [In practice, for most conventional fluids, when the Reynolds number is sufficiently large] . In these cases the discontinuities are "smoothed out" within regions whose dimensions normal to the ideal discontinuity surface are comparatively small. [In practice, of the order of the square root of $Re^{-1}$]. Within these regions (so called, dissipative regions) and only within them, the component of the diffusive fluxes normal to the ideal discontinuity is of the same order as the main component of the convective flux and there is a marked production of the entropy.
- Vortex surfaces for ideal fluids are often unstable.

II) $m = \rho V_n \neq 0$: the mass flux across the discontinuity surface is different from zero.

In this case, equations (17.2) give, for any

$$\delta H = 0$$

$$m\,\delta V_n + \delta p = 0$$

$$\delta \underline{V}_t = 0 \qquad (17.4)$$

$$m\,\delta s = \dot{s}_s \geq 0$$

These types of discontinuities are called "shock discontinuities" (see, however, later on).

- The total enthalpy is continuous (i.e. is the same on both sides) as consequence of the energy conservation.
- The tangential velocity is always continuous (as a consequence of the balance of tangential momentum)
- Upon eqs. $(17.3)_{2,3}$ :

$$\delta h = -\delta\left(\frac{V_n^2}{2}\right) \qquad (17.5)$$

$$\delta p = -m\,\delta V_n$$

Hence, since $h$ and $p$ are two independent state parameters, such a discontinuity exists if and only if the normal component of the velocity is discontinuous. [We cannot yet say, however, whether $V_n$ decreases or increases when crossing the discontinuity]. Conversely, when $\delta V_n \neq 0$ the thermodynamic states on the two sides of the discontinuity are necessarily different. It then follows that, necessarily, $\delta s \neq 0$. But if $\delta s$ is different from zero, the postulate of positive entropy production requires that it be positive. It is the very requirement $\delta s > 0$ which, as it will be seen, allows one to conclude when (i.e.for what type of fluids) $V_n$ decreases in crossing the discontinuity (so that the discontinuity itself is truly a "shock"). This part of the analysis will be carried out in subsequent lectures.

It is seen that, even in the assumption that all dissipative fluxes are negligible (i.e. with the ideal fluid model) one arrives at a discontinuity for which the contribution of the entropy production is not zero. The opposite occurred with contact and vortex surfaces. This is the reason why the mathematical abstractions of shock discontinuities proves valid also for real fluids whereas those of the other discontinuities do not. Further studies show that the actual "thickness' of a shock is of the order of the free molecular path. Within this thickness there is a smooth transition from values of the flow variables on one side of the shock to those on the other side with an intense entropy production. The mathematical abstraction of a shock discontinuity is adequate in the continuum regime provided one is not interested in knowing the details of the shock structure.

From the easily proven identity:

$$\delta(fg) = f_m \delta g + g_m \delta f \tag{17.6}$$

[where $f_m = (f_+ + f_-)/2$ and similarly for $\varrho_m$] one deduces that equations (17.4) and the mass jump condition $\delta[\varrho V_n] = 0$ can also be written as (*):

$$\delta h = -V_m \, \delta V_n$$

$$\delta p = -\frac{V_m}{\upsilon_m} \, \delta V_n \qquad (17.7)$$

$$\delta \upsilon = \frac{\upsilon_m}{V_m} \, \delta V_n$$

where $V_m = \frac{V_{n+} + V_{n-}}{2}$ is the arithmetic mean of the normal component of the velocity, $\upsilon = \frac{1}{\varrho}$ is the specific volume and $\upsilon_m = \frac{\upsilon_+ + \upsilon_-}{2}$ its mean value.

Equations (17.7) contain the two kinematical quantities $V_m$ and $\delta V_n$. Upon their elimination one obtains the relation:

$$\delta h = \upsilon_m \delta p \qquad (17.8)$$

---

(*) Since, by definition, $\varrho \upsilon = 1$ it follows from eq. (17.5) that

$$\delta \upsilon = -\frac{\upsilon_m}{\varrho_m} \, \delta \varrho$$

From the relation, $\delta(\varrho V_n) = 0$ it follows that:

$$\varrho_m \, \delta V_n + V_m \delta \varrho = 0$$

from which, on account of eq. (17.6), eq. (17.7) is obtained. Equation $(17.7)_2$ follows from eq. $(17.4)_2$ and from the easily proven identity $m = \frac{V_m}{\upsilon_m}$.

involving only thermodynamic variables. If subscripts (1) and (2) denote values of the two sides of schock discontinuity the equation (17.8) can also be written as :

$$h_1 - h_2 + \frac{(v_1 + v_2)}{2}(P_2 - P_1) = 0 \qquad (17.9)$$

This relation is known as the Rankine-Hugoniot relation. Its properties will be discussed in subsequent lectures.

A similar relation in terms of the internal energy is readily dreived since $h = u + pv$ so that $\delta h = \delta u + p_m \delta v + v_m \delta p$. Hence:

$$\delta u = -P_m \delta v$$

## 18. BOUNDARY CONDITIONS

The evaluation of a particular flow field entails the solution of the basic system of differential equations. To do this one needs to assign a number of boundary and/or initial conditions.

The number and types of conditions to be assigned in order to guarantee the existence and unicity of the solution and its continuous dependence on the prescribed data depend on the particular problem to be solved, i.e. on the type of differ- ntial equations which rule the problem itself. This aspect of the question will be taken up in subsequent lectures. It will be pointed out here that there is an intimately close relationship between the mathematical and physical aspects of the problem at hand and many pitfalls will be avoided if one always keeps in mind the physical aspect of the problem.

One such "physical" aspect to be accounted for it is the experimentally evident fact that for real fluids (i.e. fluids with non-vanishing viscosity) in the continuum regime there can be no slip of the fluid in contact with a solid surface. In other words, the relative velocity between the fluid and the solid surface must vanish at the surface itself. However, much as in the case of the already discussed contact and/or vortex discontinuities, when a suitably defined Reynolds number is sufficiently large one may, as a first approximation, replace the physical non-slip condition with the (weaker) condition expressing the fact that the solid surface is a stream surface (i.e. one prescribes that only the component of the relative velocity normal to the solid surface vanishes, when the solid surface is impermeable). As seen, this is tantamount to consider the fluid ideal and implies that the solid surface is also a vortex surface (i.e. a surface on which the vorticity is infinite). The assumed ideality of the fluid is clearly not consistent with the physical reality and will therefore lead to a number of so called "paradoxes". It can be proved, however, that the "ideal fluid" solution is still valid throughout with the exception of a thin layer, adherent to the solid surface, the so-called boundary layer. The thickness of this boundary layer is essentially the maximum thickness through which the vorticity can diffuse. Then, whenever the appropriate Reynolds number is sufficiently large one may practically solve the flow field in two subsequent steps. First one makes the ideal fluid assumption and determines the flow field without imposing the non-slip condition. It can be shown that the pressure distribution on the body derived under the ideal fluid assumption is, to within acceptable limits of accuracy, still valid for the non-ideal case. Using this pressure distribution as "input" one then evaluates the flow field within the boundary layer region. This two-step procedure in the solution of the flow fields for a sufficiently large Reynolds num-ber is often adequate. It poses, however, some often subtle and complex mathematical problems of the type known as "singular per-turbation problems".

## 19. PROPERTIES OF FLOW FIELDS AT A LARGE REYNOLDS NUMBER

The Reynolds number is defined by:

$$R_e = \frac{\varrho V L}{\mu} = \frac{V L}{\nu}$$

where $\nu = \frac{\mu}{\varrho}$ is the kinematic viscosity (which, for air in standard conditions is equal to 0.14 cm /sec). Whenever the quantities which enter the definition of $R_e$ are chosen as to be representative and characteristic of the problem being investigated, the

Reynolds number acquires important physical interpretations. Two of them are relevant to the subject course: the "dynamic" and the "entropic" interpretation.

According to the first one, $Re^{-1}$ gives a measure of the relative importance of viscous forces (i.e. forces arising from the dissipative parts of the stress tensor) compared with inertia forces. [Equivalently, $Re^{-1}$ measures the relative importance of dissipative momentum fluxes compared with the convection momentum flux].

According to the second one, $Re^{-1}$ gives a measure of the efficiencey of the entropy production associated with the dissipative part of the stress tensor, in the sense that, for sufficiently large Reynolds number there is not "sufficient time" for the natural cause of entropy production (viscosity) to produce its effects.

Closely related to the Reynolds number is the Peclet number

$$P_e = P_r \, Re$$

with the Prandtl number $P_r$ defined as:

$$P_r = \frac{c_p \mu}{\lambda} = \frac{\nu}{K}$$

where $K = (\lambda / c_p \varrho)$ is the thermal diffusivity.

The Prandtl number depends only on the fluid and on its state and expresses the ratio between the diffusivities of momentum and internal energy. For gases is of the order one; thus the two diffusivities are of the same order (i.e. the characteristic times associated with the diffusions of momentum and internal energy in gases are of the same order).

Consequently, for gases the Peclet number is of the same order as the Reynolds number.

The inverse of the Peclet number can be interpreted as giving a measure of:

a) the relative importance of diffusion of entropy compared to its convection;

b) the efficiency of the entropy production associated with heat conduction within the fluid.

Since $P_e = O(Re)$ for gases, it follows that, whenever $Re^{-1} \ll 1$ gas behaves totally as an ideal fluid, namely:

- Its stress tensor reduces to the isotropic tensor $p\,\underline{\underline{U}}$ (i.e. can "communicate" only normal, non dissipative, stresses)
- it cannot "conduct heat" (i.e. it "diffuses" its internal energy or its entropy)
- it cannot produce entropy

In the definition of the Reynolds number there appear a length which, as said, must be characteristic of the problem being studied in order for the above mentioned interpretations and conclusions following thereupon to apply. It follows that the hypothesis $Re^{-1} \ll 1$ can never be valid throughout the region of interest when there are either solid boundaries or contact and/or vortex surfaces. Nonetheless, as said, the solution obtained under the assumption $Re^{-1} \ll 1$ throughout can still be used everywhere except within the dissipative regions which surround either the solid boundaries or the discontinuity surface. It is this remark (which we have still to prove) which justifies the study of the flow fields of ideal fluids.

Some of the most important properties of the flow of ideal fluids will be discussed in what follows.

When the fluid behaves as an ideal fluid the conservations of mass, momentum (in the hypothesis of negligible gravitational effects) and energy read:

$$\varrho \frac{D\upsilon}{Dt} = \underline{\nabla} \cdot \underline{V}$$

$$\varrho \frac{D\underline{V}}{Dt} + \underline{\nabla} p = 0 \qquad (19.1)$$

$$\varrho \frac{D}{Dt}\left(h + \frac{V^2}{2}\right) - \frac{\partial p}{\partial t} = 0$$

since, as said, all dissipative fluxes are negligible.

The above system of equations is closed when one prescribes the fundamental thermodynamic relation (in terms of enthalpy, for instance)

$$h = h(s, p) \qquad (19.2)$$

For continuous motion a suitable combination of

equations (19.1) leads, when accounting for the Gibbs relation:

$$dh = Tds + v dp \tag{19.3}$$

to the entropy balance equation.

This equation however can be written down directly upon considering that, in the ideal fluid approximation (i.e. in the approximation $Re^{-1} \ll 1$ ) both the diffusion and production of entropy are negligible. It follows then that:

$$\varrho \frac{Ds}{Dt} = 0 \tag{19.4}$$

i.e. the entropy of each particle is conserved. When such is the case the motion is said to be <u>isentropic</u>. In general, however, the particles need not have the same entropy. If the flow is continuous (i.e. if the streamlines do not cross any shock surface) and if it originates from a uniform region then, upon the obvious initial conditions, the solution of eq. (19.4) is $s$= con-stant throughout. In this case (i.e. when all particle have and maintain the same entropy) the flow field is said to be <u>homoentropic</u>.

A similar terminology is used in connection with the total enthalpy

$$H = h + \frac{V^2}{2} \tag{19.5}$$

The flow is isoenthalpic if $H$=const along each streamline (but differs from one streamline to another) and <u>homo-</u>

enthalpic if H=const. throughout the flow field. Eq. (19.1) shows that the flow is certainly isoenthalpic if it is stationary (in G ). An isoenthalpic flow is certainly homoenthalpic if it originates from a uniform region, no matter whether or not shock discontinuities are present. This is so because, as said, the total enthalpy is continuous across a shock.

The quantity $\frac{D\underline{V}}{Dt}$ is nothing but the acceleration $\underline{a}$ of the elementary particle. It can also be expressed as:

$$\underline{a} = \frac{D\underline{V}}{Dt} = \frac{\partial \underline{V}}{\partial t} + \underline{\nabla}\left(\frac{V^2}{2}\right) + \left(\underline{\nabla} \wedge \underline{V}\right) \wedge \underline{V} \tag{19.6}$$

The force per unit mass $(-\underline{\nabla} P/\rho)$ acting on the particle is certainly conservative if the flow is incompressible ($\rho$ =const). In general, it will consist of two parts, one conservative and the other not. Since it is of purely thermodynamic nature, its expression must be obtained from the fundamental relation (19.3) (with $v=1/\rho$) one indeed obtains:

$$-\frac{\underline{\nabla} P}{\rho} = T\underline{\nabla}s - \underline{\nabla}h \tag{19.7}$$

Hence, by combining equations (19.6) and (19.7) the momentum equation can also be formulated as:

$$\frac{\partial \underline{V}}{\partial t} + \underline{\nabla} H + \left(\underline{\nabla} \wedge \underline{V}\right) \wedge \underline{V} = T\underline{\nabla}s \tag{19.8}$$

a relation known as Crocco's equation.

Consider steady motions $\left(\frac{\partial}{\partial t} = 0\right)$ in G originating from an uniform region. As said, in this case the motion is certainly homoenthalpic and equation (19.8) reduces to:

$$(\underline{\nabla} \wedge \underline{V}) \wedge \underline{V} = T\,\underline{\nabla} s \tag{19.9}$$

Two main cases may be considered:

i) The flow is continuous throughout (i.e. no shocks are present) Then, necessarily, it is homoentropic and consequently (aside from the case of no particular interest here that $\underline{\nabla} \wedge \underline{V}$ is parallel to $\underline{V}$) it is also necessarily irrotational (i.e. $\underline{\nabla} \wedge \underline{V} = 0$).

ii) The flow is not continuous and there is a curved shock. Since as it will be seen behind a curved shock $\underline{\nabla} s \neq 0$ the flow behind such a shock will be rotational even if it were irrotational in front of the shock.

As seen, whenever one deals with a steady flow originating from a uniform region such a flow will certainly be homoentripic and irrotational (in the ideal fluid approximation) at least up to the first possible shock discontinuity.

Steady, irrotational, homoentropic flows constitute an important class of flow fields.

This class is also a comparatively simple one.

Indeed: a) the homoentropic character reduces the independent thermodynamic field variables to only one; b) the irrotational character reduces the kinematic field variable to a single scalar variable, the so-called velocity potential $\varphi$ defined by:

$$\underline{V} = \underline{\nabla}\varphi \tag{19.10}$$

The basic system of equation can be readily reduced to a single scalar equation for the potential $\varphi$ .Multiply scalarly the momentum equation by $\underline{V}$ to obtain, on account of the stationarity of the flow:

$$\varrho\,\frac{D\,V^2/2}{Dt} + \frac{Dp}{Dt} = 0 \tag{19.11}$$

On the other hand, since $\frac{Ds}{Dt} = 0$ it is:

$$\frac{Dp}{Dt} = \left(\frac{\partial p}{\partial v}\right)\frac{Dv}{Dt} = -\frac{a^2}{v^2}\,\frac{DV}{Dt} \tag{19.12}$$

where $a^2$ denotes the square of the Laplacian characteristic speed (speed of sound). If one now accounts for the continuity equation, equation (19.11) becomes:

$$\frac{D\,V^2/2}{Dt} - a^2\,\underline{\nabla}\cdot\underline{V} = 0 \tag{19.13}$$

or, in terms of the velocity potential defined by equation (19.10):

$$(\underline{\nabla}\varphi)\cdot\underline{\nabla}\left(\frac{\underline{\nabla}\varphi}{2}\right)^2 - a^2\Delta_2\varphi = 0 \tag{19.14}$$

where $\Delta_2\varphi$ is the Laplacian of $\varphi$ . This equation is the required potential equation since $a^2$ can be expressed in terms of $\varphi$ through the energy equation. For perfect gas with constant specific heats one gets:

$$\frac{a^2}{\gamma-1} + \frac{(\underline{\nabla}\varphi)^2}{2} = H = \text{const.} \tag{19.15}$$

Particular cases are:

i) Incompressible flows. In this case $a^2 = \infty$ (in practice the flow can be considered incompressible when the Mach number $M = V/a$ is of the order $0{,}2 \div 0{,}3$) and equation (19.14) reduces to the Laplace equation:

$$\Delta_2\varphi = 0$$

ii) Linearized flows when the obstacle around which the fluid flows is sufficiently thin and the angle of attack is sufficiently small one can let:

$$\varphi = V_\infty x + \varphi_1$$

where $V_\infty$ is the free stream velocity directed along the $x$-axis, and $\varphi$ , the perturbation potential, is such that its powers greater than the first one can be neglected. To within this approximation equation (19.14) becomes

$$M_\infty^2 \frac{\partial^2\varphi_1}{\partial x} - \Delta_2\varphi_1 = 0$$

where $M_\infty^2 = V^2/a^2$ is the free stream Mach number.

For plane flows, for examples, one has:

$$\varphi_{1yy} + (1 - M_\infty^2)\varphi_{1xx} = 0$$

It is seen that this equation is hyperbolic for $M_\infty^2 > 1$ (supersonic free stream) and elliptic for $M_\infty^2 < 1$ (subsonic free stream).

## REFERENCES

1 R.Monti, L.G.Napolitano : "Numerical computations for non-equilibrium $H_2 - O_2$ rocket nozzle flow" A.R.S. Report, ELDO Future Programme Preliminary Study, N.31, Part II.

2 L.G.Napolitano : "Théorie des petites perturbations pour des écoulements hors d'équilibre" Fasc.I Thermodynamique, Cours de Troisième Cycle de Mécanique des Fluides, 1966-67 Faculté des Sciences de Paris, Gauthier-Villard; 1970.

3 S.R.De Groot and P.Mazur : "Non Equilibrium Thermodynamics" North Holland Publ. Co. Amsterdam, 1962.

4 C.E.Treanor and P.V.Marrone : "Effect of Dissociation on the Rate of Vibrational Relaxation" The Physics of Fluids vol. 5 No. 9 Sept. 1962.

5 R.Monti : "Influence of vibrational excitation on the thermodynamic properties of a reacting mixture" (In Italian) (Paper presented at the XIX Congresso Nazionale A.T.I., Siena (Italy), 7, 11 Oct. 1964).

6 P.Hammerling, J.D.Teare and B.Kivel : "Theory of Radiation from Luminous Shock Waves in Nitrogen" Phys. Fluids 2, 422 (1959).

7 R.Monti, L.G.Napolitano :"Research on Thermodynamics and dynamics of Gas Mixture with two rate processes" Interim Report, Office of Naval Research, Contract ONR-4544(00).

8 W.H.Dorrance : "Viscous Hypersonic Flow", McGraw Hill Co.

## CONTENTS

**O.M. BELOTSERKOVSKII**

*Computing Center – Academy of Sciences*
*Moscow, USSR*

# METHODS OF COMPUTATIONAL GASDYNAMICS

## Introduction

1. At present, specialists of applied sciences are confronted with various kinds of practical problems whose successful and accurate solution , in most cases, may be attained only by numerical methods with the aid of computers. Certainly, it does not mean that analytical methods which permit us to find the solution in the "closed" form will not be developed. Nevertheless, it is absolutely clear that the range of problems permitting such an approach to their solution is rather narrow, therefore, the development of general numerical algorithms for the investigation of problems of mathematical physics is important.

It is especially urgent in continuum mechanics (gas dynamics, theory of elasticity, etc.), which is accounted for by some circumstances.

1) Difficulties of carrying out the experiment. In studying the phenomena taking place, for example, at hypersonic flight velocities, the resulting high temperatures give rise to the effects of dissociation, ionization in the flow and, in a number of cases, even to "luminescence" of a gas. In these cases it is enormously difficult to simulate the experiment in the laboratory, since for the similarity between the natural environment and the conditions of the experiment it is

not sufficient to satisfy the classical criteria of similarity, i.e. the equality of the Mach and Reynolds numbers.
The equality of absolute pressures and absolute temperatures is also required, which is only possible if the sizes of the model and the real object are equal. All this involves numerous technical difficulties and is related to high cost of the experiment, to say nothing of the fact, that the results of the tests in many cases are rather scarce.

However, the importance of the experiment, in principle, must not be underestimated. The experiment is always the basis of investigation confirming (or rejecting) the scheme and the solution with some theoretical approach.

2) The complexity of the equations considered. Deep penetration of the numerical methods into continuum mechanics is also explained by the fact that equations of aerodynamics, of gas dynamics, of theory of elasticity represent the most complicated (as compared to other branches of mathematics) system of partial differential equations. In a general case, this is a nonlinear system of mixed type with the unknown form of the conversion surface (where the equations change their type) and with "movable boundaries", i.e. the boundary conditions are given on surfaces or lines which, in turn, are determined by calculations. Moreover, the range of the unknown functions is so wide that ordinary methods of analytical investigation (linearization of equations, series expansion, introduc-

tion of a small parameter etc.) do not fit for the derivation of the full solution of the problem for this general case.

It should be noted, that in solving complicated problems on electronic computers the preliminary analytical investigation of a problem may be of great aid and sometimes this investigation is simply decisive for the successful realization of the numerical algorithm.

In the long run, the smoothness of the functions represented determines the success of the use of some algorithm with the least expenditure of the computer time. Therefore, the choice of independent variables, various forms of writing down the initial system of equations (which may be mathematically equivalent, but not of equal value from the point of view of their approximate representation), the use of exact integrals of the system, the determination of directions along which the representation of functions is given, the structure of the calculation network - all these factors play an important role in developing the numerical algorithm.

3) Let us dwell on one more peculiarity of algorithms used for solving concrete problems of continuum mechanics. As is known, the numerical methods nowadays have found a wide practical application in design offices and research Institutes. Substantial progress in the exploration of cosmos, in the optimum control of vehicles, in the choice of rational configurations of vehicles and so on are, to a consider-

able extent, due to serial calculations and the use of scientific information obtained in this way. The volume of information obtained by means of the calculation is far more complete and substantially cheaper than the corresponding experimental investigations if the problem is correctly formulated, well modelled and algorithmically rational. However, a wide application of the numerical methods for practical purposes requires sufficent simplicity and reliability.

Thus, on the one hand, one has to deal with rather complicated mathematical problems, on the other hand, it is necessary to develop rather simple and reliable numerical methods permitting us to carry out serial calculations at project Institutes and design offices.

Later, the consideration will be restricted mainly, to the numerical methods for solving gas dynamics problems. In this respect, the topic of the lectures is somewhat narrower than one can expect from the title.

Note, in the first place, that for most problems in gas dynamics, not only any mathematical theorems of existence and uniqueness have been proved, but very often there is no confidence that such theorems can be derived. As a rule, the very mathematical formulation of the problem is not strictly given and only the physical treatment is presented, which is far from being one and the same thing. The mathematical difficulties of the investigation of such types of problems are related to nonlinearity of equations, as well as to a great number of independent variables.

No better is the state of affairs with the methods of solving gas dynamics equations. So far, the investigations related to the possibility of realization of the algorithm, its convergence to the unknown solution, its stability have rigorously been performed only for linear systems, and, in a number of cases, only for equations with constant coefficients. When confronted with the necessity of solving a problem, the mathematician has to use the known algoritms and to develop new methods without rigorous mathematical basis for their applicability. It does not mean, that such a situation is very much different from that in any other new field. In science, as well as in mathematics, one can find many examples when new ideas and concepts originated and were successfully used without solid basis which appeared later. Of course, it does not mean, that when developing new numerical algorithms, one must act at random without thinking about the accurate formulation of the problem or without comprehending its physical meaning. This way inevitably leads to numerous mistakes, moreover, to the waste of time; and, especially, the experience without being theoretically interpreted does not give the foundation for further development of the method.

We want to draw your attention to this rather clear question only because there is still an opinion, that the main thing is to write down differential equations and all the rest reduces to a trivial substitution of derivatives by differences and to programming, to which sometimes too much importance is attached. In this connection, it is reasonable to formulate the main stage of the numerical solution to a mechanical or physical problem with the aid of computers in the following way:

1) the construction of a physical model and the mathematical statement of a problem;

2) the development of a numerical algorithm and its theoretical interpretation;

3) programming (manual or automatic) and the formal adjustment of the program;

4) the methodical adjustment of the algorithm, i.e. the test of its operation by concrete problems; the elimination of the drawbacks found and the experimental investigation of the algorithm;

5) serial calculations, accumulation of the experience, the estimation of effectiveness and the range of applicability of the algorithm.

At all stages, the mathematical theory, the physical and numerical experiment with the aid of computers are used jointly and consistently. The way it is realized at each stage may be illustrated by solving concrete problems, which will be done below. Therefore, we shall make only some common observations.

The main principle of using mathematical results is that the conditions providing the solution of a problem for more simple and special cases must be fulfilled for more common and complicated problems. Parallel to this, the consideration of the physical meaning of the phenomenon gives a qualitative picture with the help of which the statement of the problem is checked and defined more exactly. Ultimately, the final experimental test allows us to judge about the correctness of the assumption made and to give the estimation of the algorithm and the solution derived, for example, its accuracy. It should be noted, that the estimation of the accuracy of the numerical solution of the formulated differential problem must be done purely mathematically, without using the results of the physical experiment. The latter may be used for qualitative comparison , while the quantitative comparison between the calculation and the experiment must provide information of how closely the physical model used approaches the natural environment.

2. In exact sciences there may arise many important problems, the investigation of which is tied to the solution of a system of non-linear partial differential equations. Gas dynamics is one of these sciences, and, furthermore, it includes many problems with discontinuous solutions.

The construction of reasonably accurate solutions of the exact equations of gasdynamics in the general case has become possible only with the aid of numerical methods, exploit ing the advantages of high-speed electronic digital computing machines. The requirements of technology have called for an intensive development of numerical methods and their application to the solution of a wide variety of gas dynamics problems. Scientists and research engineers in the area of gas dynamics have contributed significantly to the development of modern numerical methods of solving systems of non-linear partial differential equations.

There exist four universal numerical methods which are applicable to the solution of non-linear partial differential equations.

I. Method of Finite Differences. This method is the most highly developed of the four at the present time and is widely applied to the solution of both linear and non-linear equations of the hyperbolic, elliptic and parabolic types. The region of integration is subdivided into a network of computational cells by a generally fixed orthogonal mesh. Derivatives of functions in the various directions are replaced by finite differences of one form or another; usually, a so-called implicit difference scheme is applied to the integration of the equations. This results in the solution, at each step of the procedure, of a system of linear algebraic equations involving perhaps several hundred unknowns.

Finite difference schemes are often used for solving unsteady gas dynamics equations. Lagrangian and Eulerian approaches are widely used here. In the first case, where the coordinate network is related to the liquid particles the structure of the flow is better defined and one succeeds in constructing rather accurate numerical schemes for flows with comparatively small relative displacements. In the second case, when the calculational network is fixed over space, the schemes are used for constructing flows with large deformation. In recent time, the approaches mentioned here have found a wide application also for the calculation by stabilization of steady flows.

II. Method of Integral Relations. In this method, which is a generalization of the well-known method of straight lines, the region of integration is subdivided into strips by a series of curves, the shape of which is determined by the form of boundaries of the region. The system of partial differential equations written in divergence form is integrated across these strips, the functions occurring in the integrands being replaced by known interpolation functions. The resulting approximate system of ordinary differential equations is integrated numerically. The method of integral relations, like the method of finite differences, is applicable to equations of various types.

III. Method of Characteristics. This method is applied only to the solution of equations of hyperbolic type. The solution, in this case, is computed with the aid of a network of characteristic lines, which is constructed in the course of the computation. Actually, the method of characteristics is a difference method of integrating systems of hyperbolic equations on the characteristic calculational network and is mainly used for detailed description of flows. Its distinguishing feature as compared to other difference methods is the minimum utilization of interpolation operators and associated maximum proximity of the region of influence of the difference scheme as well as the region of influence of the system of differential equations. The smoothing of the profiles inherent in the difference schemes, with fixed network is minimum here, since the calculational network used in the method of characteristics is constructed exactly with the region of influence of the system taken into account. Irregularity (nonconservativeness) of the calculational network should be referred to the drawbacks of the method of characteristics. It is possible to develop a technique, based on this method, in which the calculations are carried out in layers bounded by fixed lines. The method of characteristics permits one accurately to determine the point of origin of secondary shock waves within the field of flow, as the result of intersection of characteristics of one family. On the other hand, if a large number of such shock waves are developed, difficulties are encountered in their calculation. Accordingly, the method of characteristics is most expediently applied to hyperbolic problems in which the number of discontinuities is small (for example, problems concerning steady supersonic gas flow).

## IV. "Particle-in-Cell" Method (PIC).

In certain respects, the Harlow PIC method incorporates the advantages of the Lagrangian and Eulerian approaches. The range of solution here is separated by the fixed (Eulerian) calculation network; however, the continuous medium is interpreted by a discrete model, i.e. the population of particles of fixed mass (Lagrangian network of particles) which move across the Eulerian network of cells is considered. The particles are used for determining parameters of the liquid itself (mass, energy, velocity), whereas the Eulerian network is employed for determining parameters of the field (pressure, density, temperature).

The PIC method allows us to investigate complex phenomena of multicomponent media in dynamics, because particles carefully "watch" free surfaces, lines of separation of the media, and so on. However, due to a discrete representation of the continuous medium (the finite number of particles in a cell) calculational instability (fluctuations) often arises here, the calculation of rarefied regions is also difficult, and so on. Limitations in power of modern computers do not permit a considerable increase in the quantity of particles.

For problems in gas dynamics in the presence of uniform medium, it seems more reasonable to use here the concept of continuity considering the mass flow across the boundaries of Eulerian cells instead of "particles".

A large number of different problems in gas dynamics have been solved with the aid of the above numerical meth ods. Admittedly, these were applied principally to the solution of systems of partial differential equations with two or three independent variables. It should be noted that, in solving multidimensional problems, the most expedient method may prove to be a combination of the various numerical methods. For example,the calculation of the three-dimensional supersonic flow about bodies of revolution at an angle of attack can be carried out by a combination of the method of integral relations and the method of characteristics, finite differences and PIC methods etc.

The aim of the present lectures is to describe briefly the application of some of these numerical methods to problems of gas dynamics.

We shall concern here the fundamental principles of constructing the numerical schemes by the method of integral relations, the network -characteristic approach and the "large particles" method. The application of these schemes as well as finite-difference methods for the calculation of subsonic, tran sonic and supersonic flows of a gas, problems of the boundary layer, motion of bodies at an angle of attack, etc. will be considered.

The work was done in the Computing Center of the Academy of Sciences of the USSR and the Moscow Physico Technical Institute. All necessary details and detailed calculations of the methods can be found in the papers referred to.

## NUMERICAL METHOD SURVEY

### 1. Steady-State Schemes

In determining the steady aerodynamic characteristics of bodies (especially when electronic computers of average power were employed) we made wide use of the following methods for solving steady gas dynamics equations, the method of integral relations (m.i.r.), the method of characteristics (m.ch.) and some finite difference schemes (e.g.,schemes with "artificial viscosity", and others). We wish to consider especially problems in which different discontinuities and singularities are given beforehand, together with some associated boundary conditions; the solutions being carried out in regions where functions vary continuously.

As is known , three different schemes of the method of integral relations were developed for the determination of flow in the region of a blunt nose, namely, using an approximation for the initial functions across the shock layer (Scheme I), along it (Scheme II) or in both directions (Scheme III). As a result, the boundary value problem was solved for an approximate system of ordinary differential (algebraic) equations. To solve the three-dimensional problems, some additional trigonometric approximation in the circumferential coordinate was introduced [21, 22] . For different flow conditions and different body shapes one of the schemes of the method of integral relations has been found applicable; they are widely used in our country as well as abroad.

The main advantage of these schemes is that, by means of different transformations, one succeeds eventually in approximating functions (or groups of functions) with comparatively weak variations. It allows us to obtain reliable results and a high degree of accuracy with a comparatively small number of interpolation nodes (usually 3 - 4 calculated points were used).

The choice of the independent variables, the form of the initial system of equations of motion (that is, the introduction of the integrals into the initial system, the use of the divergent form of the laws of conservation and others), the use of conservation schemes, the approximation of the integrals, etc., are all of great importance in writing the numerical algorithm using m.i.r., and hence in producing results.

The main difficulty in carrying out the schemes of m.i.r. is the solution of many parameter boundary problems for the approximating system of equations. This was overcome by means of different iteration schemes. Moreover, these schemes were used in transonic regions mainly for bodies of a comparatively simple form, while when dealing with a supersonic zone one had to adopt another algorithm.

In calculating supersonic flow the 2 and 3 dimensional schemes of the method of characteristics by P.I. Chushkin, K.M. Magomedov and their coworkers were used [18-20, 23-25]. As is known, having written down the initial form of the system in characteristic variables, one requires approximation of ordinary derivatives only. Using a fixed linear computational network, we get a system of finite difference equations with its several advantages.

With the help of the above mentioned approaches, a large number of gas dynamics problems have been solved, namely, ideal gas flows with chemical reactions and radiation, transonic and three-dimensional flows, as well as viscous flows. In most cases sufficiently steady and reliable results were obtained, which were in perfect agreement with experiment. However, these approaches to the solution of the steady-state equation may be successfully used only for problems in which there are no singularities, discontinuities, intersections and interactions. The application of these approaches is difficult for bodies of complex form with a large number of discontinuities. Besides, a single algorithm for the calculation of different types of flow is preferable.

## 2. Unsteady-State schemes

The next step in the evolution of numerical meth

ods, which was motivated by urgent practical needs and aided by the availability of electronic computers, was the development of nonsteady schemes and the use of the stability method for the solution of steady-state aerodynamic problems. We tried to keep to the general principles and ideas of the m.i.r. and m.ch. in approximating the nonsteady equations with respect to space variables. The divergent or characteristic forms of the initial equations were used, the same calculation networks were employed, etc.

In this way the nonsteady Schemes II and III of the method of integral relations and the network-characteristic method were developed [25] . These allow us to consider rather complicated types of flow with a single algorithm. It is natural that the problems of stability and the attainment of steady-state solutions should become crucial. They require some specific technique such as the introduction of artificial viscosity into the initial system, and of dissipation terms into the difference equations. In a number of cases the accuracy of the results obtained is less than in the steady-state methods, but these approaches enabled us to consider new classes of problems; for example, the determination of the aerodynamic characteristics of three-dimensional flow for specific configurations, the calculations of viscous transonic flows, and others.

## 3. "Large Particles" Method

Finally, in the third stage of development it seemed reasonable and advantageous to introduce the elements of the Harlow "particle-in-cell" method [26] into the algorithms. At first, only the equation of continuity is represented as the mass flow across the Euler cell, using the simplest finite difference or integral approximation along the coordinates.

Thus, the modified method of "large particles" [27, 28] came into existence, which (again by means of the stability process) allowed us to consider from one point of view such a complicated task as, for example, the subsonic, transonic, and supersonic flow past a flat-nosed body in two dimensions or with axial symmetry. Such an approach is used in calculating viscous flows, and it may permit us to study the characteristics of separated flows [28] .

Thus, this review is concerned with the development of a definite class of numerical schemes used for the estimation of the aerodynamic characteristics of vehicles. It should be stressed that the development of the numerical schemes mentioned above is determined by the improvement and extension of the ways of solving the boundary value problems for the corresponding approximating equations; by the consideration of a new, wider class of problems; by the development and improvement of the elec tronic computers, machine languages, input and output arrangements and so on. At the same time the algorithms are made according to principles of the method of integral relations, the method of characteristics and other approaches. We shall not dwell on these methods in this paper, as the details are described in the articles referred to.

## 1. Method of Finite Differences

### 1.1. On the Application of the Method to Problems in Gas Dynamics

The method of finite differences is, currently, the most widely used numerical method for the solution of non-linear systems of partial differential equations. This method, stemming from the well known work by R. Courant, K.O. Friedrichs and H. Lewy (1928) [1] is presented in an extensive series of papers. The method has also received considerable attention in the Soviet Union. A number of questions related to the numerical solution of the equations of gas dynamics with the aid of the method of finite differences are investigated in an article by S.K. Godunov and K.A. Semendyaev (1962).

As is well known, it often becomes necessary in problems of gas dynamics to deal with various types of singularities (dicontinuities) within the region of integration. The system of gas dynamics equations does not in this case have a continuous solution.

Whenever the number of such singularities is small, and it is possible to formulate for them certain boundary conditions, the problem reduces to the determination of continuous solutions in regions bounded by such singular surfaces. In constructing the numerical solution to this class of problems by the method of finite differences a fixed computational network is generally employed characterized by constant values on the coordinate lines.

On the other hand, for problems with movable singularities, it is expedient to utilize a movable network connected to these singularities and to avoid the introduction of artificial independent variables for which these networks become coordinate lines. An example of the construction of this kind of network will be presented below when describing the finite-difference method of computing discontinuous solutions to the equations of gas dynamics.

The choice of a difference scheme approximating the differential equations in a stable manner is generally not difficult in these cases. Most commonly, an implicit two-layer difference scheme is employed, in which the choice of step length may be made independently for each of the variables in accordance with the requirements of accuracy and stability.

The following problems, investigated in the present survey, belong to that class of gas-dynamics problems in which discontinuities are isolated and smooth solutions constructed by the numerical method of finite differences: the three-dimensional supersonic flow about bodies and the unsteady gas flow in a spherical blast with back pressure.

When the number of discontinuities within the re

gion of integration becomes large, it is difficult to keep track of them all. In unsteady problems the method of finite differences is also complicated by the need to calculate moving lines of discontinuity, the motion of which is not known beforehand. In order to cope with these difficulties inherent in finite difference methods, special numerical schemes of "continuous calculation" were introduced, which permit one to carry out the calculations without regard for the discontinuities.

Two approaches are adopted in constructing difference schemes of this type.

The first approach proposed, by J. Neumann and R. Richtmeyer (1950, 1959) , eliminates the discontinuities by introducing into the basic differential equations a "pseudo-viscosity". For sufficiently small coefficients of viscosity the solutions of such transformed equations will be continuous, while the discontinuities will be replaced by regions where the variables change rapidly. The limiting solution to the equations with viscosity, when the coefficient of viscosity tends towards zero, yields a generalized solution of the original system of differential equations. The existence of a unique generalized solution to the equations of gas dynamics with arbitrary equation of state remains an open question at the present time.

While calculating gas movement the shock waves appearing may be removed by the introduction of viscosity in the initial equations of gas dynamics. With sufficiently small viscosity the solution of the equation will be continuous but in the vicinity of a shock wave the velocity and pressure gradients will increase more abruptly the less the viscosity is.

Let us consider the simplest wave equation

$$\frac{\partial u}{\partial t} + \frac{\partial F(u)}{\partial x} = 0 \qquad (1.1.1)$$

Viscosity may be introduced into the main differential equation

$$\frac{\partial u_\varepsilon}{\partial t} + \frac{\partial F(u_\varepsilon)}{\partial x} = \varepsilon \frac{\partial}{\partial x} \Psi \left[\frac{\partial u_\varepsilon}{\partial x}\right] \qquad (1.1.2)$$

where $\Psi$ is a monotonous function of its argument , so that $\Psi(0) = 0$ .

The solution of equation (1.1.2.) will be already continuous and ordinary stable methods may be applied. It is possible to show, that $u_\varepsilon(x,t) \to u(x,t)$ at $\varepsilon \to 0$ (if the initial conditions for $u_\varepsilon$ and $u$ coincide).

By the numerical solution of the equation (1.1.2) the coefficient of viscosity $\varepsilon$ may be diminished together with the decrease of space and time intervals.

Replacing equation (1.1.2.) by a finite-difference relation (in which $h$ is the net step according to space, $\tau$ according to time, $n$-the number of the coordinate $t$, $m$ are the numbers of the coordinates X )

$$u_{m,n+1} - u_{m,n} + \frac{\tau}{2h}\left[F(u_{m+1,n}) + F(u_{m-1,n})\right] =$$

(1.1.3)

$$= A_{m,n}(u_{m+1,n} - u_{m,n}) - A_{m-1,n}(u_{m,n} - u_{m-1,n})$$

it is necessary to choose the coefficient $A_{m,n}$ in such a way as to approximate the operator (1.1.1.) on smooth intervals of the solution by the operator (1.1.3.) with the second order of accuracy.

In the neighborhood of the discontinuities eqs. (1.1.3.) must be of the first order of accuracy. We may obtain this by submitting the coefficient $A_{m,n}$ to a definite dependence upon their values of $u$ in two neighboring point of a network $X=(M-1)h, mh$ . For linear equations we may show the convergence of the method if eq.(1.1.3.) provides a stable calculation. But the experience also confirms the convergence in the non-linear case.

The given method for an equation of the type (1.1.1.) as well as for systems of equations (in the last case $u$ and $F(u)$ may be regarded as vectors, and the coefficients $A_{m,n}$ as matrices), and also for the equations with two space independent variables of the type

(1.1.4) $$\frac{\partial u}{\partial t} + \frac{\partial F(u)}{\partial x} + \frac{\partial G(u)}{\partial y} = 0$$

($u$, $F$ and $G$, $n$-dimensional vectors).

The construction of approximate numerical schemes for equations with viscosity is extremely difficult, since in the regions of rapidly varying gradients the accuracy of the calculations decreases and it becomes necessary to introduce a large number of nodes in the network. In addition, the same difficulties arise here as in the numerical computation of smooth solutions - the possibility of accumulation of singularities. The question of the character of such an accumulation and of the possibility of continuing the solution, must be considered.

The other approach to the determination of discontinuous solutions, developed by P.D. Lax (1954) and S.K. Godunov (1959, 1962) [5,6], consists in the construction of special difference schemes reflecting the conservation laws at the discontinuities. In the realm of gas dynamics these are the laws of conservation of mass, momentum and energy. If, as the step length reduces to zero, the finite difference scheme tends towards the exact relationship on the discontinuities, the solution of the corresponding difference equations will tend to the required discontinuous solution.

For instance, the system of gas dynamics equations in Lagrangian coordinates

$$\frac{\partial u}{\partial t} + \frac{\partial p(e,v)}{\partial x} = 0 \qquad (1.1.5.a)$$

$$\frac{\partial v}{\partial t} - \frac{\partial u}{\partial x} = 0$$

(1.1.5.b)

$$\frac{\partial}{\partial t}\left(e + \frac{u^2}{2}\right) + \frac{\partial pu}{\partial x} = 0$$

correspond to integral relations

$$\oint u dx - p dt = 0$$

(1.1.6)

$$\oint v dx + u dt = 0$$

$$\oint \left(e + \frac{u^2}{2}\right) dx - pu dt = 0$$

One of the difference schemes, providing the solu tion of the integral relations (1.1.6) may be written down in the form

$$u_{m+1/2}(t+\tau) = u_{m+1/2}(t) - \frac{\tau}{h}\left(p_{m+1} - p_m\right)$$

(1.1.7)

$$v_{m+1/2}(t+\tau) = v_{m+1/2}(t) + \frac{\tau}{h}\left(u_{m+1} - u_m\right)$$

$$\left(e + \frac{u^2}{2}\right)_{m+1/2}(t+\tau) = \left(e + \frac{u^2}{2}\right)_{m+1/2}(t) - \frac{\tau}{h}\left(p_{m+1}u_{m+1} - p_m u_m\right)$$

while the speed and pressure at the bounds of the layer $u$ and $p$ may be obtained from relations (exact or approximate) on discontinuities. This method was proposed by S.K. Godunov.

In this chapter, schemes for continuous calculation will be demonstrated in the course of describing difference methods for finding discontinuous solutions to the equations of gas dynamics, and in investigating the numerical calculation of the interaction between unsteady shock waves and obstacles.

## 1.2. Three dimensional supersonic flow about bodies [2].

Let us now consider the three-dimensional problems of calculating a purely supersonic region of a perfect gas flow about bodies at an angle of attack. The surface of the body is presumed smooth everywhere with the exception of the apex. For the numerical solution of this problem K.I. Babenko and G.I. Voskresenkii (1961) [2] constructed a three-dimensional difference scheme, which permitted calculations to be made in that region where the flow is supersonic and continuous. Following the authors, we shall describe their method for solving this problem.

Here it is convenient to use the coordinates $x$, $\xi = (r - r_b)/(r_s - r_b)$, $\Psi$ where $x, r, \Psi$ are cylindrical coordinates, $r = r_b(x, \Psi)$ is the given equation of the body contour, $r = r_s(x, \Psi)$ is the desired equation of the wave. The x-axis is chosen so that the projection of the velocity on it will be greater than the local sonic velocity.

In the given case of supersonic flow, the system of gas dynamics equations will be an x-hyperbolic system, which for a perfect gas can be presented in the following vector form:

$$\frac{\partial \vec{x}}{\partial x} + \vec{A}\,\frac{\partial \vec{x}}{\partial \xi} + \vec{B}\,\frac{\partial \vec{x}}{\partial \Psi} + \vec{C} = 0\,, \tag{1.2.1.}$$

where $\vec{x}$ is a five-dimensional vector, the components of which are the three velocity components, the pressure and the density; $\vec{A}, \vec{B}, \vec{C}$ are matrices dependent on the above five functions. The boundary conditions for this system (1.2.1.) on the body ( $\xi = 0$ ) are solid surface conditions, while on the shock wave ( $\xi = 1$ ) they consist of the well known conservation laws relating to the fluxes of mass, momentum and energy. In addition, it is assumed that at some $x = x_0$ initial values are known for all of the unknown functions.

Let us now replace the system (1.2.1.) by a system of difference equations. The given region is subdivided by a three-dimensional network with nodes

$$(h_1 n\,,\; h_2 m\,,\; h_3 \ell)\,,\; n = n_0\,,\; n_0 + 1\,, \ldots\,;\; m = 0,1,\ldots M$$

$\ell = 0,1,2 \ldots \mathcal{L}$ in which $Mh_2 = 1, \mathcal{L} h_3 = 2\pi$. Values of the function $f$ at the node point $(h_1 n, h_2 m, h_3 \ell)$ will be written $f^{n}_{m,\ell}$ (Fig. 1.) (see page 249). We will be considering, in addition to nodes with integral coordinates $(n, m, \ell)$ points lying just midway between these values (values of functions at these points will be distinguished in an analogous fashion).

The following scheme will be used to replace partial derivatives by difference expressions

$$\left(\frac{\partial f}{\partial x}\right)^{n+\frac{1}{2}}_{m+\frac{1}{2},l} = \frac{1}{2h_1}\left(f^{n+1}_{m+1,l} - f^{n}_{m+1,l} + f^{n+1}_{m,l} - f^{n}_{m,l}\right),$$

$$\left(\frac{\partial f}{\partial \xi}\right)^{n+\frac{1}{2}}_{m+\frac{1}{2},l} = \frac{1}{h_2}\left[\lambda\left(f^{n+1}_{m+1,l} - f^{n+1}_{m,l}\right) + \beta\left(f^{n}_{m+1,l} - f^{n}_{m,l}\right)\right], \tag{1.2.2}$$

$$\left(\frac{\partial f}{\partial \Psi}\right)^{n+\frac{1}{2}}_{m+\frac{1}{2},l} = \frac{1}{4h_3}\left[\gamma\left(f^{n+1}_{m+1,l+1} - f^{n+1}_{m+1,l-1} + f^{n+1}_{m,l+1} - f^{n+1}_{m,l-1}\right) + \right.$$

$$\left. + \delta\left(f^{n}_{m+1,l+1} - f^{n}_{m+1,l-1} + f^{n}_{m,l+1} - f^{n}_{m,l-1}\right)\right],$$

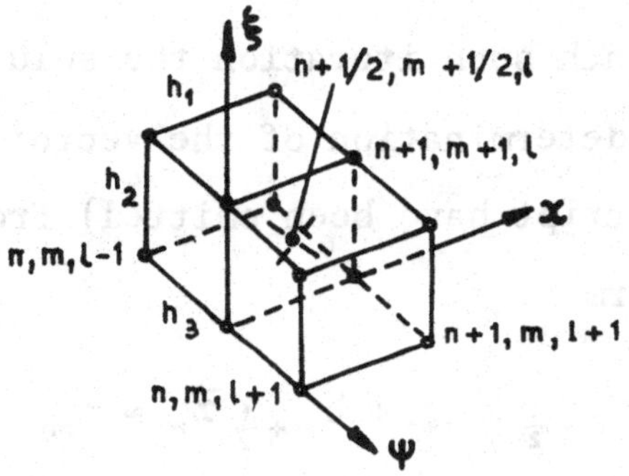


Fig. 1

where $\lambda, \beta, \gamma, \delta$ are positive numbers restricted by the conditions

$$\lambda + \beta = 1 \quad , \quad \gamma + \delta = 1$$

In satisfying the system (1.2.1) at the points

$\left[\left(n+\frac{1}{2}\right)h_1, \left(m+\frac{1}{2}\right)h_2, \ell h_3\right]$, We obtain a non-linear system of difference equations in the following form:

$$\vec{x}^{\,n+1}_{m+1,\ell} + \vec{x}^{\,n+1}_{m,\ell} + 2\frac{h_1}{h_2}\lambda\vec{A}^{\,n+\frac{1}{2}}_{m+\frac{1}{2},\ell}\left(\vec{x}_{m+1,\ell} - \vec{x}_{m,\ell}\right)^{n+1} +$$

$$+\frac{h_1}{2h_3}\gamma\vec{B}^{\,n+\frac{1}{2}}_{m+\frac{1}{2},\ell}\left(\vec{x}_{m+1,\ell+1} - \vec{x}_{m+1,\ell-1} + \vec{x}_{m,\ell+1} - \vec{x}_{m,\ell-1}\right)^{n+1} =$$

(1.2.3)

$$= -2h_1\vec{C}^{\,n+\frac{1}{2}}_{m+\frac{1}{2},\ell} + \vec{x}^{\,n}_{m+1,\ell} + \vec{x}^{\,n}_{m,\ell} - 2\frac{h_1}{h_2}\beta\vec{A}^{\,n+\frac{1}{2}}_{m+\frac{1}{2},\ell}\left(\vec{x}_{m+1,\ell} - \vec{x}_{m,\ell}\right)^{n} -$$

$$-\frac{h_1}{2h_3}\delta\vec{B}^{\,n+\frac{1}{2}}_{m+\frac{1}{2},\ell}\left(\vec{x}_{m+1,\ell+1} - \vec{x}_{m+1,\ell-1} + \vec{x}_{m,\ell+1} - \vec{x}_{m,\ell-1}\right)^{n},$$

which at any $n$ is solved by an iterative method.

At each such iteration the solution of the problem reduces to the determination of the vector $\vec{z}$ (the superscript and the subscript have been omitted) from the difference equations of the form

(1.2.4) $$a_{m+\frac{1}{2}}\vec{z}_{m+1} + b_{m+\frac{1}{2}}\vec{z}_m = f_m$$

with known boundary conditions on the body $(m = 0)$ and on the shock wave $(m = M)$. This problem is solved with the aid of the so-called "marching method". We shall now explain the essence of this method.

With the aid of the equation (1.2.3.) one can obtain for the vector the relationship

(1.2.5) $$\left(\vec{\mu}_m \cdot \vec{z}_m\right) = g_m$$

written in the form of the scalar product of the unit four-dimensional vector $\vec{\mu}_m$ and the vector $\vec{z}_m$. At $m=0$ the values $\vec{\mu}_0$ and $g_0$ are easily obtained from the boundary conditions on the body. At other values of $m$ it is possible to determine the $\vec{\mu}_m$ and $g_m$ with the aid of recurrence formulas, from eq. (1.2.4).

$$\vec{\mu}_{m+1} = \frac{a'_{m+\frac{1}{2}}\left(b^{-1}_{m+\frac{1}{2}}\right)\vec{\mu}_m}{\left|a'_{m+\frac{1}{2}}\left(b^{-1}_{m+\frac{1}{2}}\right)'\vec{\mu}_m\right|}, \quad g_{m+1} = \frac{-g_m + \left[f_{m+\frac{1}{2}}\left(b^{-1}_{m+\frac{1}{2}}\right)\vec{\mu}_m\right]}{\left|a'_{m+\frac{1}{2}}\left(b^{-1}_{m+\frac{1}{2}}\right)'\vec{\mu}_m\right|}. \quad (1.2.6)$$

Here the primes on the matrices denote transposition, while

$$|\vec{F}| = \sqrt{(\vec{F}\cdot\vec{F})}.$$

Rewriting relationship (1.2.5.) on the shock wave $m=M$ one obtains

$$(\vec{\mu}_\mu \cdot \vec{z}) = g_\mu. \quad (1.2.7)$$

This equation, together with the known boundary conditions on the shock wave, is sufficient to determine the vectors $\vec{z}_\mu$. Subsequently, combining eqs. (1.2.4.) and (1.2.5.) one obtains the vector $\vec{z}_m$ in terms of the known vector $\vec{z}_{m+1}$.

Next, it is necessary to determine initial values in the plane $x=x_0$. If the body has a pointed head, then for a sufficiently small region in the neighborhood of the point, the surface of the body is replaced by a cone, the calculated flow about which yields the necessary initial data. Since the flow in the neighborhood of a cone at an angle of attack is of similarity type it is possible to start with some initial data (the calculation is most conveniently begun with the known solution for the cone at zero angle of attack, the latter then gradually increased), and subsequently determine the conical flow of interest, using a fairly large number of steps in $x$ for tabulation of the solution. For bodies with blunt headforms the initial data would stem from the numerical solution of the corresponding flow about the nose.

With the aid of this method, K.I. Babenko and G.P. Voskresenskii [2] calculated the flow about a circular cone and a body of revolution at an angle of attack. Some of the results of the calculations are presented here in a series of graphs. All of the variables involved are in dimensionless form; the velocity components $u, v, w$ in the direction $x, r, \Psi$, respectively, are related to the critical sonic velocity, $a_{cr}$, the density $\rho$ to the density of the incident flow $\rho_\infty$, the pressure $p$ to the quantity $\rho_\infty a_{cr}^2$.

The results of calculations of supersonic flow about a body of revolution having parabolic generators and a conical nose for various Mach numbers $M_\infty$, angles of attack $\alpha$, and fineness of the computational network (these data are given in the tabular form) are given in Fig. 2., showing the traces on a vertical plane of the shock waves. For the same body with $M_\infty = 3 \cdot 5$ and $\alpha = 20^\circ$ the distribution of the velocity components $u, w$, the density $\rho$ and the pressure $p$ on the body surface (solid line) and shock wave (dotted line) in a series of meridian planes $\Psi =$ constant are presented in Fig. 3.

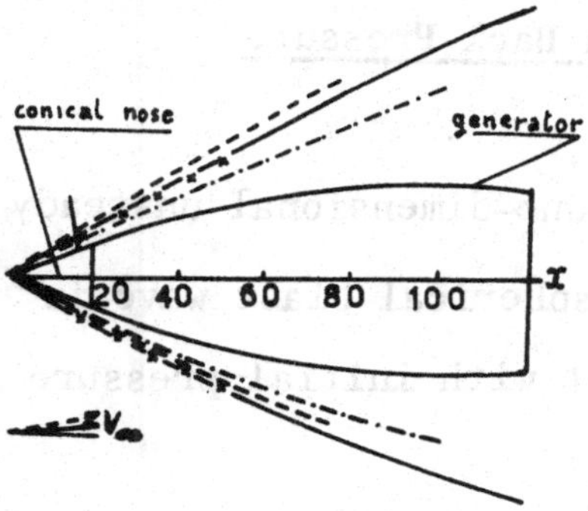

| Notation | $\alpha$ | $h_2$ | $h_3$ | $M_\infty$ |
|---|---|---|---|---|
| —————— | 5° | 0,05 | $\pi/8$ | 3,5 |
| – – – – | 10° | 0,05 | $\pi/8$ | 3,5 |
| –·—·—·— | 5° | 0,05 | $\pi/8$ | 6 |
| × | 5° | 0,05 | $\pi/8$ | 3,5 |

Fig. 2

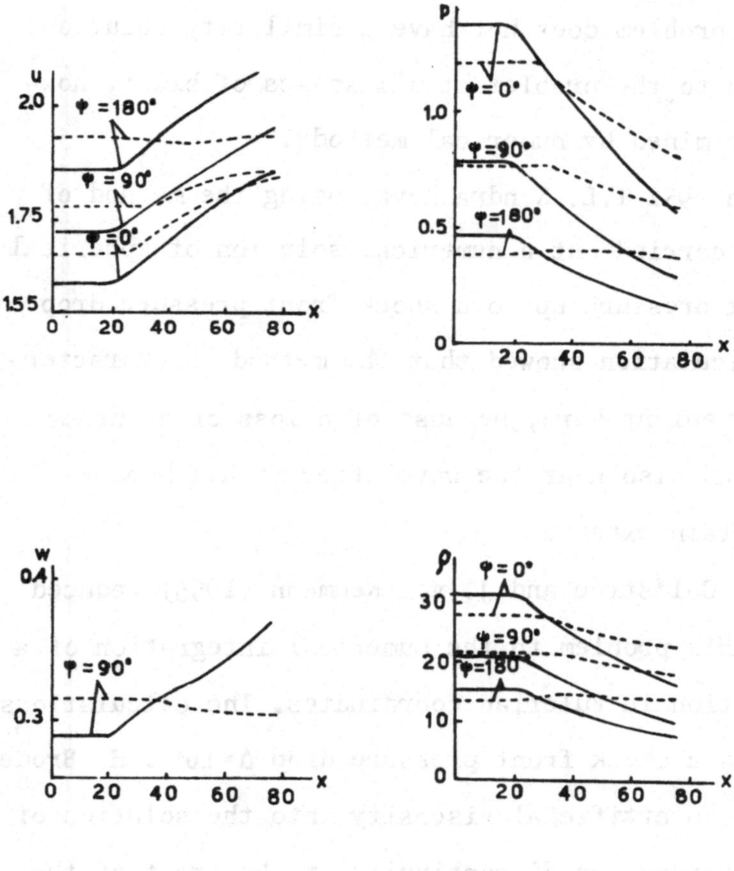

Fig. 3

## 1.3. Calculation of Spherical Blast with Back Pressure

We shall now examine the one-dimensional unsteady problem concerning the propagation of a spherical blast wave in a gas of infinite extent initially at rest with initial pressure $p_\infty$ and initial density $\rho_\infty$.

As is well known, when the back pressure $p_\infty$ is not negligible compared with the pressure at the front of the blast wave, this problem does not have a similarity solution. An exact solution to the problem at all stages of blast, however, can be determined by numerical methods.

In 1952 I.L. Kondrasheva, using the method of characteristics, carried out a numerical solution of spherical blast with a back pressure up to a shock front pressure drop $\Delta p/p_\infty = 2.3$. This calculation showed that the method of characteristics, was inconvenient here, because of a loss of accuracy near the center and also near the wave after it had become weakened to a certain extent.

H. Goldstine and J. von Neumann (1955) reduced the solution of this problem to the numerical integration of a second order equation in Eulerian coordinates. The calculations were carried up to a shock front pressure drop $\Delta = 1.017$. H. Brode (1955) introduced an artificial viscosity into the solution of this problem, to remove the discontinuity at the front of the shock wave.

D.E. Okhotsimskii, I.L. Kondrasheva, Z.N. Vlasova, and R.K. Kazakova (1957) [3] calculated the disturbance due to a spherical blast by a finite difference method. The governing system of equations for this one-dimensional gas flow was written in Lagrangian coordinates. The resulting three-dimensional network enclosed the entire region behind the shock wave.

The calculated flow exibited the spontaneous generation of a compression phase and an expansion phase lying immediately behind the shock wave. These were generated when the pressure drop across the wave was $\Delta \approx 1.14$ . In the remainder of regions surrounding the center of the blast all of the parameters had values close to those corresponding to the undisturbed medium. In passing to each new time layer, the region of integration over the Langragian coordinate increased as the shock wave propagated outward. Consequently, the number of computational points lying in the essentially disturbed region consisting of these two phases gradually decreased. In view of this, the calculations could not be continued beyond a pressure drop $\Delta = 1.008$.

In order to determine the behavior of a spherical blast wave at large distances from the center of blast D.E. Okhotsimskii and Z.P. Vlasova (1962) [4] undertook a new calculation with initial data taken from Okhotsimskii et al. (1962) corresponding to a pressure drop $\Delta = 1.1073$ . The computational network enclosed only the disturbed region immediately behind the shock wave. The remainder of the region enclosed by the wave was assumed at rest. Let us now have a brief look at the details of the solution.

In view of the fact that in the late stages of the blast the gas dynamic variables differ only slightly in magnitude from their values in the undisturbed medium, new functions were introduced by Okhotsimskii and Vlasova (1962) representing corrections to the principal parts of the values of pressure, density and particle velocities. To calculate the coordinates of the shock fronts a linear acoustic part was extracted from the propagation velocity in just the same way.

For the numerical solution of this problem the following dimensional variables $\eta, \tau$ were introduced:

$$\eta = \frac{\bar{r} + \lambda}{\bar{r}_s + \lambda} \quad , \quad \tau = \frac{t}{t^0}$$

Here $\lambda$ is a constant, $\bar{r} = r / r^0$,

$$r^0 = \left(\frac{E_0}{p_\infty}\right)^{\frac{1}{j+1}} , \quad t^0 = r^0 \left(\frac{\rho_\infty}{p_\infty}\right)^{\frac{1}{2}}, \tag{1.3.1}$$

$t^0$ and $r^0$ are dynamic time and length, respectively, $E_0$ - the blast energy $j = 2$ for the spherical case, and the subscript s refers to conditions on the shock wave. In the $\eta, \tau$ variables the region of integration comprises a rectangle.

In the governing equations (continuity, momentum and adiabatic condition) for spherical symmetric unsteady gas flow the partial derivatives were replaced by a second order difference scheme

$$\frac{\partial f}{\partial \tau} = \frac{f_{m+1}^{n+1} + f_{m}^{n+1} - f_{m+1}^{n} - f_{m}^{n}}{2\Delta\tau},$$

$$\frac{\partial f}{\partial \eta} = \frac{f_{m+1}^{n+1} - f_{m}^{n+1} + (1-\delta)(f_{m+1}^{n} - f_{m}^{n})}{\Delta\eta},$$

where $f_m^n$ is the value of the function $f(\eta,\tau)$ at a node $(m\Delta\eta, n\Delta\tau)$ while $\delta$ is a weighting factor $\frac{1}{2} \leq \delta < 1$.On each time layer the boundary value problem for the system of difference equations is solved by a marching method, proposed by I.M. Gel'fand and O.V. Lokutsievskii. In this formulation the calculations were carried out up to a sharp front pressure drop $\Delta = 1.0018$ .

In order to carry out the calculations up to a very large time, approximate terminal relationships, obtained by S.A. Khristianovich (1956) were used, based on the assumption that at large distances from the center of blast, the difference between the pressure on the shock front and the pressure in the undisturbed medium is small. This enabled us to compute the distribution of parameters right up to the time that the shock wave profile overturns in the region of its negative phase.

We shall cite here one of the results of this calculation of spherical blast in air ( $\gamma = 1.4$) at a late stage of development. In Fig. 4 there are constructed velocity profiles (referred to the corresponding velocity directly behind the shock wave $V_s$), in the disturbed region, at various times $\tau$. The dotted

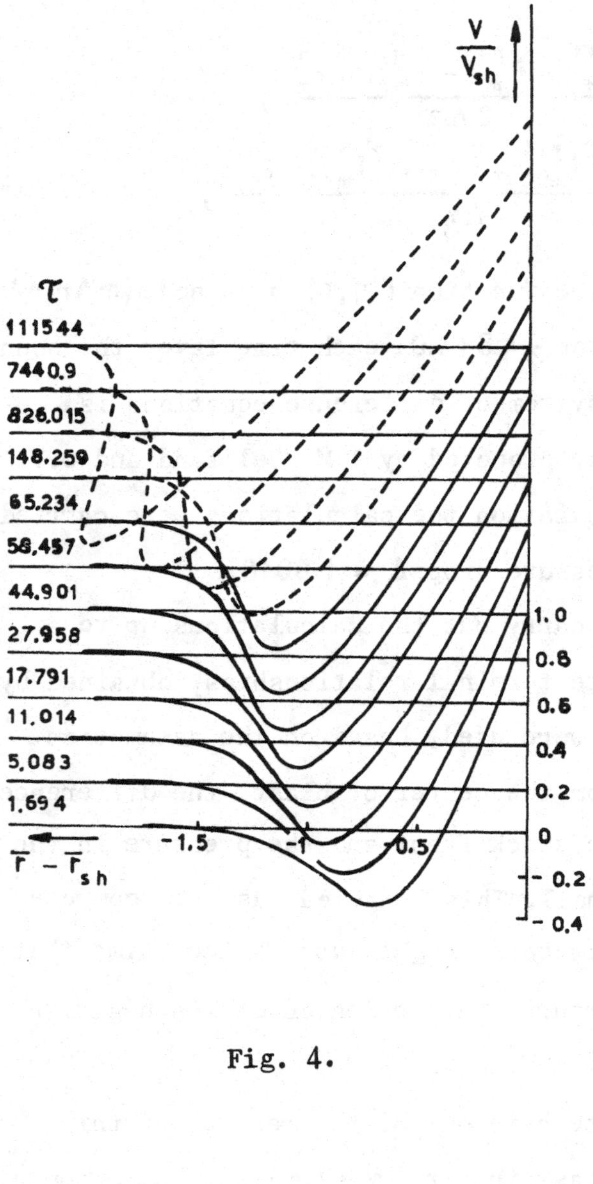


curves have been obtained by the approximate method of S.A. Khristianovich (1956). For clarity each successive curve is displaced along the coordinate axis, relative to the preceding curve. The curves show the overturning of the velocity profile and the generation of a secondary shock wave in the region of the negative phase, in the time interval $826 < \tau < 7441$.

Fig. 4.

## 1.4. Calculation of Unsteady Discontinuous Solutions

The equations of gas dynamics consist of conservation laws; consequently, they may be present in divergence form. In the one dimensional unsteady case the system of gas dynamics equations in Eulerian variables $t, x$ has the following form:

$$\begin{gathered}\frac{\partial \rho}{\partial t} + \frac{\partial \rho u}{\partial x} = 0, \\ \frac{\partial \rho u}{\partial t} + \frac{\partial (p + \rho u^2)}{\partial x} = 0, \\ \frac{\partial}{\partial t}\left[\rho\left(e + \frac{u^2}{2}\right)\right] + \frac{\partial}{\partial x}\left[\rho u\left(e + \frac{p}{\rho} + \frac{u^2}{2}\right)\right] = 0\end{gathered} \tag{1.4.1}$$

where $u$ is the velocity, $\rho$ density, $p$ pressure and $e$ the internal energy per unit mass of a gas. This system of equations is valid in most cases where the corresponding derivatives of the desired functions are continuous.

After integrating the system over a certain region bounded by a contour $L$, we obtain the integral relations

$$\begin{gathered}\oint \rho dx - \rho u dt = 0, \\ \oint \rho u dx - (p + \rho u^2) dt = 0, \\ \oint \rho\left(e + \frac{u^2}{2}\right) dx - \rho u\left(e + \frac{p}{\rho} + \frac{u^2}{2}\right) dt = 0\end{gathered} \tag{1.4.2}$$

These relations are valid for arbitrary integrable functions. They formed the basis of a finite difference scheme for calculating one-dimensional unsteady flows, developed in the Soviet Union by S.K. Godunov (1959, 1962) [5]. We shall now describe this scheme as presented in these references.

Let us choose for the contour $L$ a rectangle from the computational network, with corners $(x_m, t_n)$, $(x_{m+1}, t_n)$, $(x_{m+1}, t_{n+1})$, $(x_m, t_{n+1})$. The integral relations (1.4.2.) can then be written in the form

$$\rho_{m+\frac{1}{2}}^{n+1} = \rho_{m+\frac{1}{2}}^{n} - \frac{h_1}{h_2}\left[(RU)_{m+1}^{n+\frac{1}{2}} - (RU)_{m}^{n+\frac{1}{2}}\right],$$

$$(\rho u)_{m+\frac{1}{2}}^{n+1} = (\rho u)_{m+\frac{1}{2}}^{n} - \frac{h_1}{h_2}\left[(P+RU^2)_{m+1}^{n+\frac{1}{2}} - (P+RU^2)_{m}^{n+\frac{1}{2}}\right], \tag{1.4.3}$$

$$\left[\rho\left(e+\frac{u^2}{2}\right)\right]_{m+\frac{1}{2}}^{n+1} =$$

$$= \left[\rho\left(e+\frac{u^2}{2}\right)\right]_{m+\frac{1}{2}}^{n} - \frac{h_1}{h_2}\left[RU\left(E+\frac{P}{R}+\frac{U^2}{2}\right)\right]_{m-1}^{n+\frac{1}{2}} - \left[RU\left(E+\frac{P}{R}+\frac{U^2}{2}\right)\right]_{m}^{n+\frac{1}{2}}.$$

Here $h_1 = t_{n+1} - t_n$, $h_2 = x_{m+1} - x_m$, the index $m$ denotes a functional value at $x = x_m$ the index $n$ designates the value of a function at $t = t_n$, the index $m+\frac{1}{2}$ some average value in the interval $(x_m, x_{m+1})$ and $n+\frac{1}{2}$ some average value of velocity, density, pressure and internal energy on the boundary of the $x$ layers.

If all the gas dynamics variables were continuous, it would not be difficult to determine the average values of these quantities in terms of their value point. But since it is desirable to construct a method of calculation suitable also for application to shock waves, we shall consider each point in the region as as being a possible point of discontinuity. For a finite difference scheme this means that each node of the network must be viewed as a position of discontinuity.

Since velocities and pressures in neighboring layers $m-\frac{1}{2}$ and $m+\frac{1}{2}$ may be different, there will occur on the boundary between them (at the point $x_m$) a breakdown in continuous gas flow and waves will emanate in both directions from this point. The change in the gas dynamic variables at this centered singularity will be described by a similarity type solution of the form:

$$u = u\left(\frac{x-x_m}{t-t_0}\right), \quad \rho = \rho\left(\frac{x-x_m}{t-t_0}\right),$$

$$p = p\left(\frac{x-x_m}{t-t_0}\right), \quad e = \left(\frac{x-x_m}{t-t_0}\right).$$

It follows that the values of the functions $u, \rho, p$ and $e$ at points $x=x_m$ (i.e., the quantities $U, R, P, E$) will remain constant until the waves coming from the centered singularity at the neighboring points $x_{m-1}$ and $x_{m+1}$ arrive at the point $x_m$.

Consequently, in eqs. (1.4.3) the time step $h_1$ should not be too great, so that the values of the unknown functions at all points $x_m$ may be assumed constant. These values may be calculated in accordance with well known methods from these conditions on the discontinuity.

The three equations (1.4.3.), together with the equation of state

$$p^{n+1}_{m+\frac{1}{2}} = p\left(\rho^{n+1}_{m+\frac{1}{2}}, e^{n+1}_{m+\frac{1}{2}}\right)$$

can be used to determine the four quantities $u^{n+1}_{m+\frac{1}{2}}, \rho^{n+1}_{m+\frac{1}{2}}, p^{n+1}_{m+\frac{1}{2}}$ ,

$e^{n+1}_{m+\frac{1}{2}}$ . These magnitudes will describe approximately the state of the gas at the instant time $t^{n+1} = t^n + h_1$ .

It should be noted that, with smooth functions, this difference scheme yields an approximation to the differential equations (1.4.1.) with first order accuracy. The fact that the basis for constructing the scheme uses conservation laws, ensures that the approximate solution obtained with its aid will be close to the generalized solution of the equations of gas dynamics, even if these generalized solutions contain discontinuities.

S.K. Godunov, A.V. Zabrodin and G.P.Prokopov(1961) [6] generalized the finite difference scheme described above for calculating one-dimensional unsteady flows, to the case of two-dimensional unsteady flows of a perfect gas. In the latter, using conservation laws, one obtains integral relations in which the integrals are now taken over a close surface in three-dimensional space $(x, y, t)$ .

The work of these authors is of interest also because their numerical method may be utilized for calculations with a movable finite difference network. This is convenient when carrying out gas dynamic calculations in regions whose boundaries consist of movable lines. In using a movable network a calculation of one step in time consists of three stages. First, one determines the displacement of the boundaries of the region, in the course of one time step. After that, the motion of the network and its new position are determined. Finally, one determines the values of the several variables in all of the nodes of the new network at the new time by means of the difference scheme.

With the aid of this finite difference scheme S. K. Godunov, A.V. Zabrodin and G.P. Prokopov (1961) calculated the steady flow about a sphere in a supersonic air stream ($\gamma$ =1.2.5) with incident Mach number ($M_{\infty} = 4$).

The form of the computational network is shown in Fig. 5,the calculation is carried out in the region ABCD , where AD is the axis of symmetry, DC the contour of the body. The exterior boundary AD is prescribed in some arbitrary manner before the computations have begun. In the course of the calculations it is displaced and finally coincides with the position of the detached shock wave. The ray B is fixed, and prescribed in such a manner that the minimum region of influence will be contained within the region of computation into K sectors; the segment of each ray contained within this region is subdivided into M parts; in this manner the body contour and the shock wave sought in the solution are approximated by poligonal lines.

In Fig. 6 there are shown certain of the computed results for flow about a sphere. Fig. 6 illustrates the establishment in the course of time of the steady value of density at one of the internal cells of the network. In Fig. 7 there are constructed the flow patterns (shock wave and sonic line) and pressure distribution referred to the quantity $\rho_{\infty} V_{max}^2$ ($\rho_{\infty}$ is the density of the incident flow and $V_{max}$-the maximum adiabatic speed in the gas). On these graphs the number "2" denotes data obtained by the method described with a coarse network ($K=6, M=4$) while the number "3" refers to calculations with a fine network ($K=12, M=8$). The corresponding results of O.M. Belotserkovskii (1960), obtained by the method of integral relations, are shown for comparison on the same graphs and denoted by the number "1".

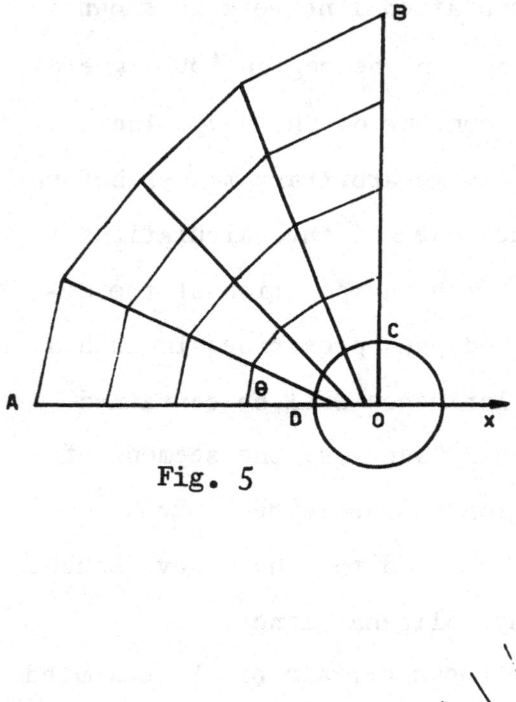


Fig. 5

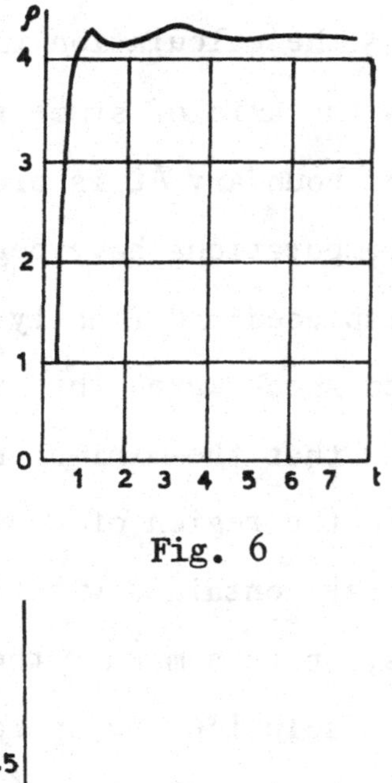


Fig. 6

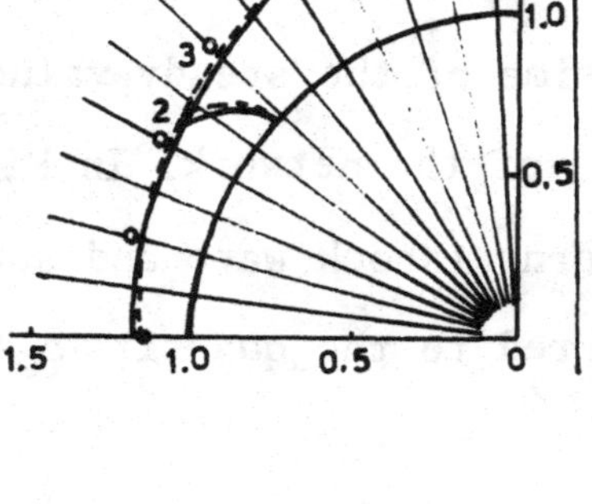


Fig. 7

## 1.5. Interaction of Shock Waves with Obstacles [7]

A numerical determination of the propagation and interaction of two-dimensional unsteady shock waves may be carried out by the method of finite differences. However, in these problems, the shock waves propagate in accordance with complicated laws unknown at the start; it is therefore expedient to use special difference schemes, permitting a continuous calculation without isolating the discontinuities. In the work of Ludloff and Friedman (1955 a, b) leading to the solution of the plane two-dimensional problem of diffraction of a shock wave by a wedge, a difference scheme of the type proposed by Lax (1954) was applied. V.V. Rusanov (1961) [7] examined the problem of the interaction between two-dimensional unsteady shock waves and obstacles of various shapes. In constructing a numerical solution he also developed a difference scheme for continuous computation. However, this has greater accuracy and is applicable to the calculation of three-dimensional axisymmetric flows of a perfect gas as well as two-dimensional flows.

We shall briefly describe V. V. Rusanov's solution [7] (1961).

In two-dimensional unsteady problems it is expedient to formulate the phenomenon in Eulerian coordinates (in spite of the fact that tangential discontinuities in density and velocity in two-dimensional problems are strongly diffused in these coordinates). The corresponding system of gas dynamic equations was written in $x, y, t$ variables, where $x, y$ are Cartesian coordinates (the $x$ axis directed along the axis of symmetry) and $t$ is the time. The equations of the problem are written in divergence form, assuring satisfaction of the conservation laws.

$$\frac{\partial f}{\partial t} + \frac{\partial F^x}{\partial x} + \frac{\partial F^y}{\partial y} + \Psi = 0 \tag{1.5.1}$$

Here $f, F^x, F^y, \Psi$ are the following four-dimensional vectors

$$f = \begin{Bmatrix} \rho \\ \chi \\ \omega \\ e \end{Bmatrix}, \quad F^x = \begin{Bmatrix} \chi \\ p + \chi u \\ \chi u \\ (e+p)u \end{Bmatrix}, \quad F^y = \begin{Bmatrix} \omega \\ \omega u \\ p + \omega v \\ (e+p)v \end{Bmatrix}, \quad \Psi = \frac{jv}{y} \begin{Bmatrix} \rho \\ \chi \\ \omega \\ e+p \end{Bmatrix}$$

In addition, $\chi = \rho u$, $\omega = \rho v$

$$e = \rho\left[\frac{(u^2+v^2)}{2} + \frac{a^2}{\gamma(\gamma-1)}\right], \quad a^2 = \frac{\gamma p}{\rho},$$

In the above formulas $u$ and $v$ are the components of the velocity $\vec{V}$ in the $x$-and $y$-directions; $p$, $\rho$ and $e$ are the pressure, density, and energy of the gas, respectively, $a$ is the sonic velocity, $\gamma$ is the adiabatic index, $j=0$ for plane flow and $j=1$ for axially symmetric flow.

We shall study the flow of a gas in regions bounded by fixed walls with rectilinear contours. The boundary conditions for the system (1.5.1.) are the solid body conditions of the wall and the symmetry condition on the axis $y = 0$.

Let us construct in $(x,y,t)$ space a rectangular network with step lengths $h_1, h_2$ and $h_3$ in the directions of the respective coordinate lines. The value of any function f at the node point of a cell with coordinates $(mh_1, \ell h_2, nh_3)$ will be designated by $f^n_{m,\ell}$.

The governing system of equations (1.5.1.) will be approximated by an elementary difference scheme which, for example, at interior points will have the form

$$f^{n+1}_{m,\ell} = f^n_{m,\ell} - h_3 \Psi^n_{m,\ell} - \frac{\lambda_1}{2}\left(F^x_{m+1,\ell} - F^x_{m-1,\ell}\right) - \frac{\lambda_2}{2}\left(F^y_{m,\ell+1} - F^y_{m,\ell-1}\right)^n + \frac{1}{2}\left[\Phi^x_{m+\frac{1}{2},\ell} - \Phi^x_{m-\frac{1}{2},\ell} + \Phi^y_{m,\ell+\frac{1}{2}} - \Phi^y_{m,\ell-\frac{1}{2}}\right],$$

where

$$\Phi^x_{m+\frac{1}{2},\ell} = \alpha^n_{m+\frac{1}{2},\ell}\left(f_{m+1,\ell} - f_{m,\ell}\right)^n, \quad \alpha^n_{m+\frac{1}{2},\ell} = \frac{1}{2}\left(\alpha_{m+1,\ell} + \alpha_{m,\ell}\right)^n,$$

$$\Phi^y_{m,\ell+\frac{1}{2}} = \beta^n_{m,\ell+\frac{1}{2}}\left(f_{m,\ell+1} - f_{m,\ell}\right)^n, \quad \beta_{m,\ell+\frac{1}{2}} = \frac{1}{2}\left(\beta_{m,\ell+1} + \beta_{m,\ell}\right)^n,$$

$$\alpha^n_{m,\ell} = \Omega\,\frac{\lambda_1^2}{\lambda}\left(V + \alpha\right)^n_{m,\ell}, \quad \beta_{m,\ell} = \Omega\,\frac{\lambda_2^2}{\lambda}\left(V + \alpha\right)^n_{m,\ell},$$

$$\lambda_1 = \frac{h_3}{h_1}, \quad \lambda_2 = \frac{h_3}{h_2}, \quad \lambda = \sqrt{\lambda_1^2 + \lambda_2^2}$$

The parameter $\Omega$ is restricted by the following stability condition

$$\left[\lambda\left(V+a\right)^n_{m,\ell}\right]^2 \leq \Omega\lambda\left(V+a\right)^n_{m,\ell} \leq 1 .$$

The fact that the gas dynamic equations (1.5.1) are written in divergence form, and the fact that the difference scheme has a special construction, permit a continuous calculation to be applied to the system. An appropriate choice of coefficients $\alpha^n_{m,\ell}$ and $\beta^n_{m,\ell}$ for the second differences ensures a greater accuracy in the computations than in Lax's scheme (1954). By means of the Fourier method used by Rusanov (1961) the stability of the given difference scheme was studied, and the permissible range of values of the coefficients $\alpha^n_{m,\ell}$ and $\beta^n_{m,\ell}$ was determined.

With the aid of the above method V.V. Rusanov carried out the computation of a number of cases of interaction of plane and axisymmetric shock waves with obstacles of various forms. All the typical problems studied are shown schematically in Fig. 8, in which the dotted line indicates the initial position of the shock wave, the arrow shows the direction of its motion and the solid line gives the distorted form of the shock wave after interaction.

Problems of the type 1 and 2 have similarity solutions (inasmuch as they lack a characteristic linear dimension), while problems of type 3 do not possess similarity solu-

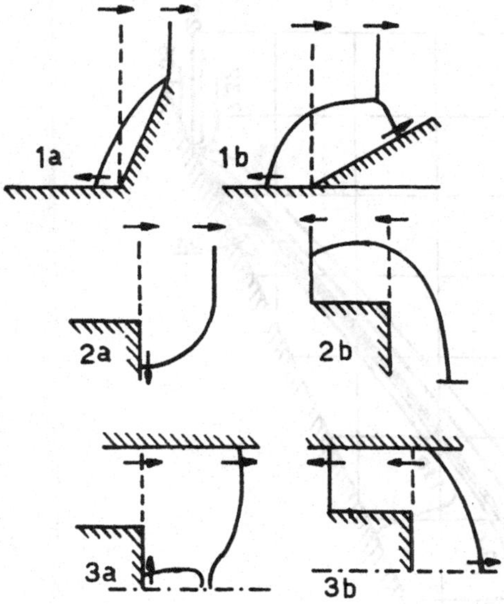

Fig. 8

tions. In similarity type problems the solution depends, in general, only upon two dimensionless variables $\xi = x/ct$ and $\eta = y/ct$ where $c$ is the velocity of the shock wave. On the other hand, when these problems are solved numerically by a scheme permitting continuous calculation the similarity characteristics are destroyed because of the smearing of the shock waves. However, at large times the distance $\Delta\xi$, over which the shock wave is smeared in the $\xi, \eta$ variables tends to 0.

This section is concluded with several illustrations of calculated flow patterns. In every example, the shock wave moves in air ($\gamma = 1{,}4$) and we take $p_- = 4p_+$, $\varrho_- = 2{,}5\varrho_+$, where the subscripts + and - designate the state of the gas ahead of and behind the original shock.

In Fig.9 there are constructed in the $\xi$, $\eta$ coordinate system lines of constant pressure $p$, referred to $p_+$, for regular reflection of a shock wave from a wedge of angle $\Theta = 65°$. The analogous graph for the Mach reflection with generation of a triple point is shown in Fig. 10 for a wedge with angle $\Theta = 30°$.

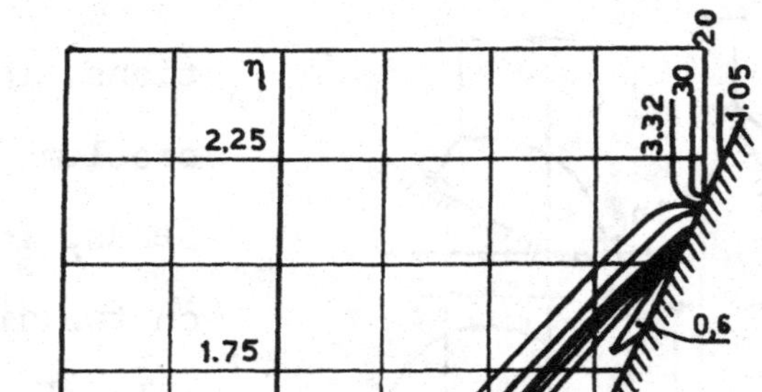
η
2,25
1.75
1.25
0.25
20
30
3.32
4.05
0,6
110
10.0
4.08
5.0
9.0
8.0
ξ
- 0.5
-0.25
0
0.25
0.5
0.75
1.0

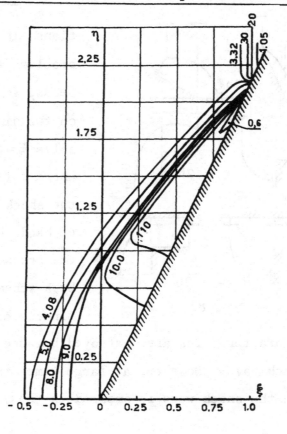

Fig. 9

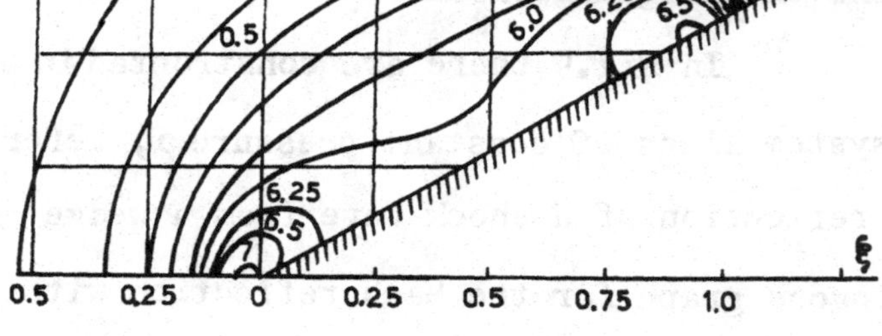
η
1.0
4.01
4.25
4.75
5.25
5.75
6.0
6.25
6.5
0.5
6.25
6.5
7
ξ
0.5
0.25
0
0.25
0.5
0.75
1.0

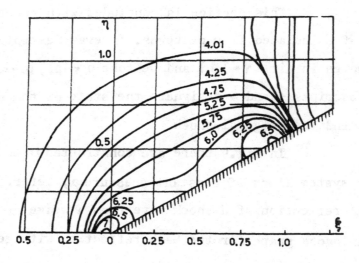

Fig. 10

## 2. Method of Integral Relations

### 2.1 Fundamental Principles of the Method of Integral Relations

The numerical method of integral relations was proposed in 1951 by A.A. Dorodnitsyn as an extension to the well-known method of straight lines for solving certain non-linear problems of aerodynamics [8], which was developed by O.M. Belotserkovskii and P.I. Chuskin for calculating potential gas flows and flows with shock waves [9]. In these methods, as has already been indicated, the integration of systems of non-linear partial differential equations reduces to the numerical solution of some approximating systems of ordinary differential equations.

In the classical method of straight lines, this is accomplished by subdividing the region of integration, usually by fixed straight lines, and replacing derivatives across these strips by finite-difference (generally linear) relationships. In the method of integral relations the strips are made, generally speaking, curvilinear, conforming to the shape of the region of integration, while across these strips integral relations are formulated, the integrand functions of which are interpolation expressions of the most general form. As a result, it is possible to obtain by the method of integral relations better accuracy with a smaller number of strips than in the method of straight lines, a factor of considerable importance in practical calculations.

Later, it was proposed a generalized method of integral relations [16] for calculating flows in a boundary layer. In this, the introduction of smoothing functions, chosen to take account of the expected behavior of the unknown function, leads to a system of ordinary differential equations in relatively smooth (as compared to the simple method of integral relations) functions. This leads to an increase in accuracy without increasing the number of strips.

The method of integral relations has the advantage that it makes use of the well-developed area of numerical solution of systems of ordinary differential equations. Furthermore, in the case of unbounded regions asymptotic methods may be applied to the solution of these equations. Programs for execution on electronic computing machines prove to be relatively simple and do not require a large memory size. On the other hand, in applying the method of integral relations difficulties are encountered when it is necessary to solve the boundary problem for an approximating system of ordinary differential equations of high order. In this case, the method proves effective if it gives sufficiently accurate results with a relatively low order of the system.

The development of the numerical method of integral relations and its applications to various problems in gas dynamics has been carried out over a period of years in the Computing center of the Academy of Sciences of USSR. The foundations of the present chapter lie in the results of these studies which, in turn, have been examined in detail in our survey [9] (Belotserkovskii and Chushkin, 1962). Other less detailed surveys of the method of integral relations can be found in the earlier published references (Dorodnitsyn, 1959, Dorodnitsyn, 1960, Katskova et al., 1958, Belotserkovskii and Chushkin, 1961).

Let us describe the fundamental premises of the numerical method of integral relations.

In this method the governing differential equations are written in divergence form. This form is suitable for presenting the differential equations of physics and mechanics extressing the laws of conservation of mass, momentum, energy, charge, etc. In the two-dimensional case, the governing differential equations in divergence form will be taken in the following general form:

$$\frac{\partial}{\partial x} P_i(x, y, u_1, \ldots, u_K) + \frac{\partial}{\partial y} Q_i(x, y, u_1, \ldots, u_K) = F_i(x, y, u_1, \ldots, u_K),$$
$$i = 1, 2, \ldots, K, \qquad (2.1.1)$$

where $x$ and $y$ are the independent variables

$u_1, \ldots, u_K$ are the unknown functions

$P_i, Q_i, F_i$ are known functions of $x, y, u_1, \ldots, u_K$.

Let the solution of the system (2.1.) be found in a region in the shape of a curvilinear quadrangle with bound aries Fig. 11.

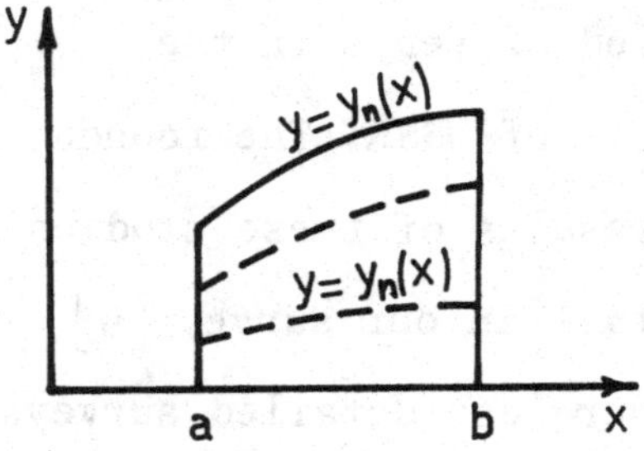


Fig. 11

$$x = a\,,\quad x = b\,,\quad y = 0\,,\quad y = \Delta(x)$$

In individual cases it may be that

$$a = -\infty\,,\quad b = +\infty\,,\quad \Delta(x) = \text{const}\,.$$

Concerning boundary conditions of system (2.1.1) we shall assume that there are in all $k$ conditions on the boundaries $x=a$ and $x=b$, and $k$ conditions on the boundaries $y=0$ and $y=\Delta(x)$. If there are singular points on a boundary the corresponding boundary conditions may be absent; they are replaced by the condition of regularity of the solution.

Multiplying each of the equations of system (2.1.1.) by some piecewise continuous function $f(y)$ and integrating with respect to $y$ from $y=0$ to $y=\Delta(x)$, we obtain integral relations for $i=1,2,\dots,k$ (the index $i$ is dropped)

$$\frac{d}{dx}\int_0^{\Delta(x)} f(y)\,P\,dy - \Delta'(x)\,f(\Delta)\,P_\Delta + f(\Delta)\,Q_\Delta - f(0)\,Q_0 -$$

$$-\int_0^{\Delta(x)} Q\,f'(y)\,dy = \int_0^{\Delta(x)} f(y)\,F\,dy. \tag{2.1.2}$$

Here

$$P_\Delta = p\left[x, y, u_1, (x,y), \dots, u_k(x,y)\right]\Big|_{y=\Delta(x)}.$$

The quantities $Q_0$ and $Q_\Delta$ are defined analogously.

Let us note that the function $f(y)$ may have a finite number of first order discontinuities. Then $\int_0^{\Delta(x)} f(y)\,Q\,dy$

should be replaced by $\int_0^{\Delta(x)} Q\, df$ .

In the method of integral relations the solution is constructed in successive approximations. Let us consider the $N$th approximation. The region of integration is divided into $N$ strips by constructing $N-1$ lines between the boundaries $y=0$ and $y=\Delta(x)$ (see Fig. 11)

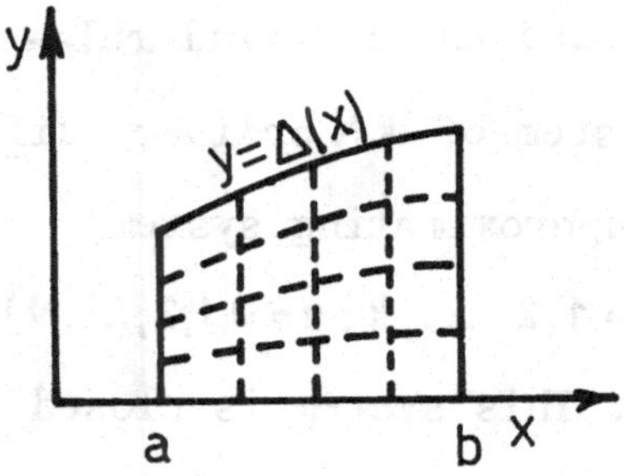

Fig. 12

$$y = y_n(x) = \frac{n}{N}\,\Delta(x) \qquad (n = 1, 2, \dots, N-1) \tag{2.1.3}$$

where $y_0 = 0$, $y_N(x) = \Delta(x)$.

The functions $P, Q, F$ will be represented by certain interpolation formulas in terms of their values $P_n, Q_n, F_n$ of the strips $y = y_n(x)$ .

Next for each value of $i$, the system of $N$ linearly independent functions $F_n(y)$ is chosen

$$\{f_n(y)\}_i \equiv \{f_1(y), f_2(y), \dots, f_N(y)\}_i \tag{2.1.4}$$

The system of these functions must be closed. Systems of functions $\{f_n(y)\}_i$ for various $i$ may coincide.

In the $N$th approximation an integral relation of the form (2.1.2) is written for each function $f_n(y)$ for the system (2.1.4). In all there will be $kN$ linearly independent integral relations. The integrand functions $P, Q,$ and $F$ are then replaced by the corresponding interpolation expressions. Then the integrals entering into the integral relations will be of the form

$$\int_0^{\Delta(x)} f_n(y)\, P\, dy \approx \Delta(x) \sum_{n=0}^{N} A_n P_n(x), \tag{2.1.5}$$

where $A_n$ are numerical coefficients, depending upon the choice of interpolation formulas and the form of the function $f_n(y)$, and represent, in essence, the generalized Coates coefficients.

Substituting eq. (2.1.5) into the integral relations of form (2.1.2) we obtain the system of $kN$ ordinary differential equations in $x$ (the so-called approximating system) for $k(N+1)$ unknown functions $u_{\nu n}(x)$ $(\nu = 1,2,\dots,k;\ n = 0,1,2,\dots,N)$ on the boundaries of the strips $y = y_n(x)$. This system is closed by a $k$ boundary conditions on the outer boundaries $y_0 = 0$ and $y_N = \Delta(x)$. The boundary conditions at $x = 0$ and $x = b$ complete the system of boundary conditions for the approximating system.

The resulting approximating system of ordinary differential equations is integrated by an approximate numerical method on an electronic digital computer.

We know that functions $P_n(x)$ enter linearly into eq. (2.1.5). As a result, the approximating system of ordinary differential equations will be linear with respect to derivatives $dP_n/dx$, and, hence, it can be easily brought to computational form by a simple solution with respect to the derivatives. Such a solution of the system can be carried out on the same electronic digital computing machine.

The functions $f_n(y)$ are chosen, generally speaking, quite arbitrarily. Let us examine the two following particular cases.

1. If one assumes for $f_n(y)$ the Dirac $\delta$ function

$$f_n(y) = \delta(y - y_n) \qquad (n = 1, 2, \ldots, N)$$

this will give the well-known method of straight lines in which derivatives with respect to $y$ are replaced by finite difference expressions corresponding to the chosen interpolation formulae.

2. If one assumes a step function

$$f_n(y) = \begin{cases} 0 & \text{for} \quad y < y_{n-1}, \\ 1 & \text{for} \quad y_{n-1} \leq y \leq y_n, \\ 0 & \text{for} \quad y > y_n, \end{cases}$$

then the simple method of integral relations is obtained. The conservation laws expressed by the system (2.1.1) will be written in this case for strips in the form of the following integral relations:

$$\frac{d}{dx}\int_{y_{n-1}}^{y_n} P\,dy - y_n' P_n + y_{n-1}' P_{n-1} + Q_n - Q_{n-1} = \int_{y_{n-1}}^{y_n} F\,dy. \qquad (2.1.6)$$

By using a scheme of this type a large group of gas dynamic problems were solved.

Yet, another form of the method of integral relations is possible. In two- dimensional problems the region of integration may be subdivided not into strips but into sub-regions (Fig.s 12, 13)

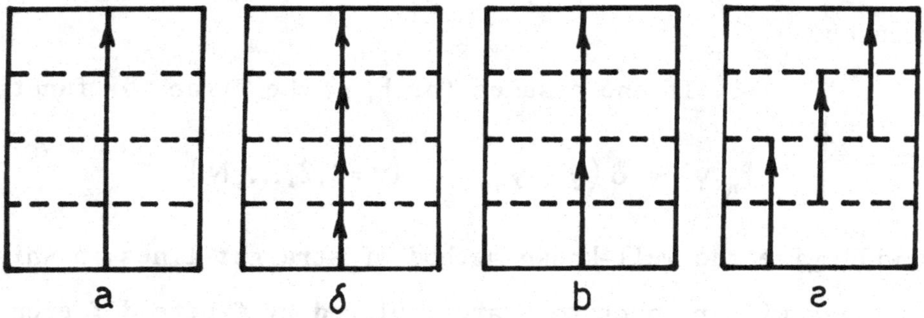


Fig. 13

and the integrations carried out in two directions, the functions being approximated in two variables. The approximating system will then become a system of non-linear algebraic or trascendental equations, which is then numerically solved on an electronic computing machine.

Finally, the method of integral relations may be generalized to the case of three-dimensional partial differential equations. Two approaches are possible here. In the first, the governing system of equations is multiplied by a function depending on two of the variables in question (the analogue of the function $f_n(y)$ in the two-dimensional case) and integrated with respect to these two variables. Applying an appropriate quadrature formula to the double intervals thus obtained leads to a system of ordinary differential equations with respect to the third variable.

In the second approach, the governing system of equations is first integrated with respect to one of the variables and the integrand functions are represented in terms of this variable by some interpolation expressions, the coefficients of which depend on the two other variables. Then the approximating system will consist of partial differential equations in these two directions and it can then be solved by the method of integral relations developed for two-dimensional problems.

We shall now make note of the characteristic properties of the method of integral relations and the advantages associated with it.

The distinguishing feature of the method of integral relations is the fact that the solution to the problem is subdivided into two separate stages. The first stage consists in the reduction of the exact system of partial differential equations into an approximating system of ordinary differential equations. The second stage, which can then be considered independently, consists in the numerical integration of this approximating system.

In this method it is, in effect, an integral which is being approximated. The accuracy of the approximation in this is increased somewhat as a result of a decrease in the coefficient of the remainder term. In addition, an integral represents a smoother function than the integrand function, thus the integral is more readily represented by a small number of interpolation nodes. Finally, the integral has a continuous representation even when the integrand function has a first order discontinuity.

In the method of integral relations the governing equations are taken in divergence form. This form is convenient and useful because in so doing integration with respect to one of the variables is carried out exactly. Furthermore, integral relations constructed in accordance with equations in divergence form and reflecting conservation equations remain valid even when crossing a surface of discontinuity. Thus, the method of integral relations may be used to carry out calculatios right across a surface of discontinuity.

We shall not dwell here on the questions of practical application of this method. Such questions (choice of independent variables and form of the governing system of equations, ways of subdividing the region into strips, direction of the approximations, choice of interpolation formulae, evaluation of accuracy, numerical solution of the approximating system of ordinary equations, etc.) are partially elucidated in Belotserkovskii and Chushkin (1962) [9], [21-22].

A large cycle of problems in gas dynamics has been solved with the aid of the method of integral relations. Computations were made principally of two-dimensional (plane, axisymmetric, conical) steady gas flows. Those studied included exterior and interior steady potential flows and problems involving the rotational flows of perfect and real gases with shock waves about bodies, the unsteady problem of gas flow in a blast with back pressure, and boundary layer flow on bodies immersed in a viscous gas. The equations describing the above gas flow were of various types: elliptic, parabolic, hyperbolic and mixed. In each of these cases, some scheme based upon the method of integral relations was used, taking into account specific characteristics of the problem.

The application of the method to the numerical solution of a series of problems in gas dynamics will be described below.

## 2.2. Potential Gas Flows [10, 11].

### 1. Subsonic Gas Flow about a Body.

The solutions to a series of problems in subsonic flows about two-dimensional axisymmetric bodies were obtained by P.I. Chushkin (1957, 1958, 1959) by the simplified method of integral relations.

In Chushkin (1957) [10] the subsonic flow in the neighborhood of ellipses and ellipsoids of revolution is examined for the symmetrical case, in which the velocity of the gas flow at infinity is parallel to the axis of the body. The problem and its solution were formulated in terms of elliptic coordinates (Fig. 14).

The equations of the problem, the equation of continuity and the equation of irrotationality have the following form:

$$\frac{\partial \chi (\sinh \xi \sin \eta)^j}{\partial \xi} + \frac{\partial \omega (\sinh \xi \sin \eta)^j}{\partial \eta} = 0$$

(2.2.1)

$$\frac{\partial \lambda}{\partial \xi} - \frac{\partial \mu}{\partial \eta} = 0$$

where

$$\chi = H_{II} \rho u , \quad \omega = H_I \rho v , \quad \lambda = H_{II} v , \quad \mu = H_I u ,$$

$$\rho = (1 - u^2 - v^2)^{\frac{1}{\gamma - 1}} , \quad H_I = H_{II} = \sqrt{\sinh^2 \xi + \sin^2 \eta} ,$$

$H_I$ and $H_{II}$ are the Lamé parameters, $u$, $v$-the velocity components in the coordinate directions referred to the maximum adiabatic velocity in the gas, $\rho$ is the density of the gas referred to the stagnation density, $\gamma$-the adiabatic index, $j = 0, 1$ for the two-dimensional axi-symmetric cases, respectively.

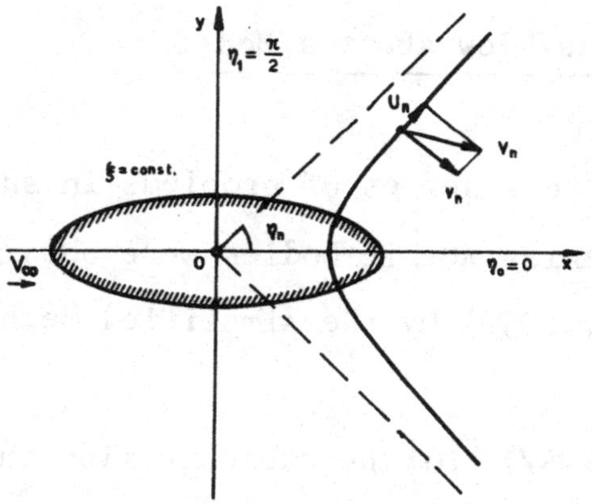


Fig. 14

If the velocity in the field of flow is everywhere subsonic, the system of equations (2.2.1) will be of elliptic type. Boundary conditions are applied at infinity where a uniform flow prevails ($u \to V_\infty \cos \eta$ , $v \to V_\infty \sin \eta$ ), and on the body where the normal component of velocity vanishes ($u = 0$).

In view of the symmetry of the problem it is sufficient to consider only one quadrant $0 \leq \eta \leq \pi/2$, where, in the Nth approximation there are traced $N-1$ hyperbolas equally spaced with respect to $\eta = \eta_n$. In the $2N$ independent integral relations formulated in accordance with the differential equations (2.2.1) the following trigonometric approximations are applied:

$$\chi = \sum_{n=0}^{N-1} a_{2n+1}(\xi)\cos(2n+1)\eta\,,$$

$$\lambda = \sum_{n=1}^{N-1} b_{2n+1}(\xi)\sin(2n+1)\eta\,, \qquad (2.2.2)$$

where $a_{2n+1}$ and $b_{2n+1}$ are linearly dependent on the values $\chi_n$ and $\lambda_n$ - on the boundaries of the strips. These approximations, arising from the known solution for the incompressible flow about an ellipse, assume rapid convergence of the solution. In addition, approximations of this form lead to a uniform flow at the singular point at infinity and in this fashion satisfy exactly the boundary conditions at this point.

The approximating system of $2N$ ordinary differential equations with respect to $\xi$ relate the velocity components $u_n$ and $v_n$ on the axes of symmetry and the hyperbolas $\eta = \eta_n$. The numerical integration of this approximating system is carried out from the body out to some sufficiently large ellipse, where the solution matches the Prandtl solution of the linearized equations for the disturbance potential. A solution of the corresponding boundary problem (with $N-1$ selected values of the function $v_n$) is not difficult to automate for execution on electronic digital computers.

In Chushkin (1957) the subsonic flow about a series of ellipses and ellipsoids of revolution was calculated for various degrees of eccentricity $\delta$ or ratio of the semi--axes. In particular, the critical Mach numbers $M_{cr}$ were determined with great accuracy for these bodies in air flow ( $\gamma = 1.4$ ). In Fig. 15 there is shown the dependence of the critical Mach number on the relative thickness of the body obtained in the various approximations ( $N = 1,2,3$ ) by the method of integral relations. These results show how well the method converges [10].

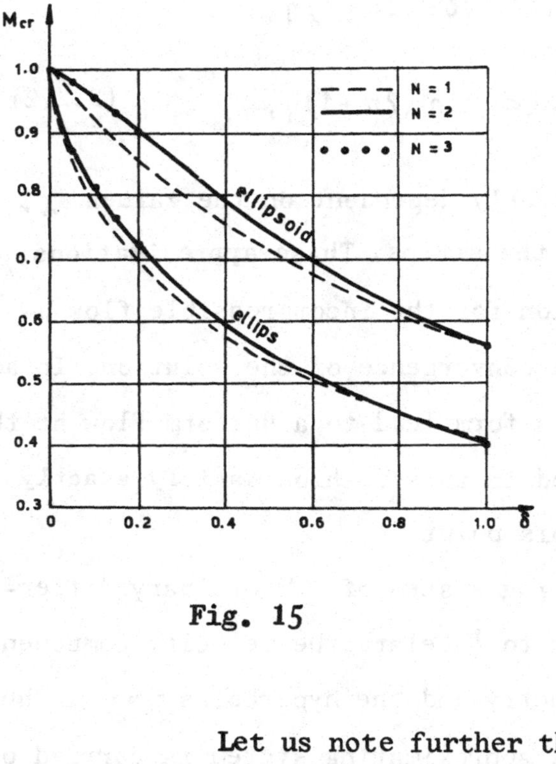


Fig. 15

Let us note further that the generalized method of integral relations with systems of functions of the form $\{\cos n\eta\}$ and $\{\sin n\eta\}$ yields for the given problem exacly the same results as with the simple method of integral relations with the approximations (2.2.2).

The subsonic gas flow about an arbitrary symmetrical profile or a body of revolution at zero angle of attack was considered in Chushkin (1958) [11]. The solution of this problem was also formulated in an elliptic coordinate system with foci in the interior of the body. In order to deal with a finite region of integration the equations of continuity and vorticity (2.2.1) are transformed to the independent variables $t = e^{-\xi}$. Within this region (in view of symmetry it is sufficient to study only the half plane $0 \leq \eta \leq \pi$), $N-1$ intermediate lines drawn between the given body contour $t = t_1(\eta)$ and the boundary $t = 0$ in the Nth approximation (Fig. 16).

$$t = t_1(\eta) = \frac{N-n+1}{N}\, t_1(\eta) \quad (n = 2, 3, \ldots, N). \tag{2.3.3}$$

Thus, in contrast to the preceding problem the approximation in the integral relations are constructed here, not with respect to the variable $\eta$ but with respect to $t$. In these approximations the character of the flow at infinity is taken into account, and so they have the form

$$F = \sum_{n=0}^{N+1+j} a_n(\eta) \left(\frac{t}{t_1}\right)^n, \tag{2.2.4}$$

where, in two-dimensional plane flow ($j = 0$) without circulation $a_1 = 0$, while in axisymmetric flow ($j = 1$) $a_1 = a_2 = 0$.

In this scheme the approximating system of ordinary differential equations contains one superfluous equation, which may be utilized for determining one of the velocity components at infinity, and for comparing this approximate calcul-

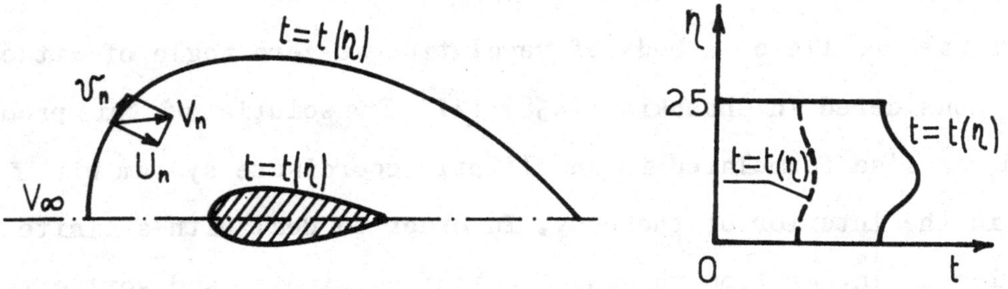


Fig. 16

ated quantity with its given value. In other words, this superfluous equation permits an evaluation to be made of the error of the approximation.

A series of examples were computed by the above method. In particular, for control purpose a calculation was made of the flow of an incompressible fluid about a symmetrical Joukowsky profile. The results obtained by the method of integral relations with $N=2$ agree very well with the exact solution. This is apparent in fig. 17, in which there is shown the distribution of velocity on a symmetrical Joukowsky profile with relative thickness $\delta = 0.13$ .

## 2. Symmetrical Flow about Bodies Moving at Sonic Velocity [12]

The method of integral relations for the solutions of this problem of mixed type was developed by P.I. Chushkin [12] (1957), who considered the sonic gas flow (incident Mach) number $M_\infty = 1$ ) about two-dimensional and axisymmetric bodies with elliptic and also arbitrary contours. The solution is constructed in special orthogonal coordinates $\xi, \eta$ in which

$$\xi = r - r_b(\Theta) , \quad \int \frac{d\eta}{r_b'(\eta)} = - \frac{1}{r} +$$

$$(2.2.5) \qquad + \int \frac{d\Theta}{r_b'(\Theta)} ,$$

where $r = r_b(\Theta)$ is the equation of the body contour in polar coordinates. Thus, the line $\xi = 0$ represents the body contour, while at infinity these coordinates blend into the polar coordinate system. For the elliptic contour this coordinate system was replaced by an ordinary elliptic coordinate system.

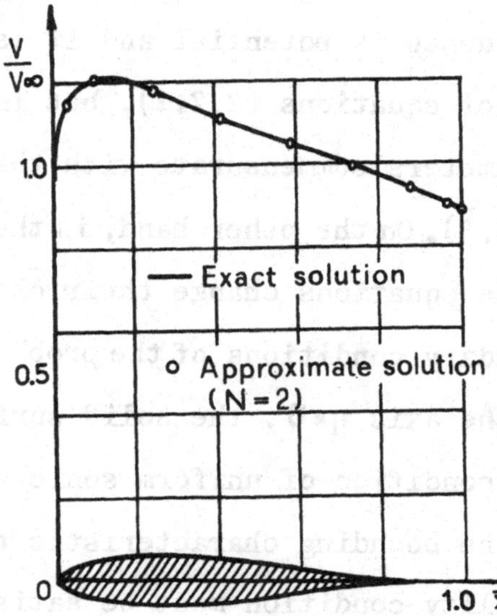


Fig. 17

It is possible in the given flow (only the upper half plane is considered) to isolate a minimal region of influence bounded by the axis $\eta = 0$, part of the body contour and the semi-infinite characteristic $\eta = \eta_1(\xi)$ of initially unknown form and tangent at infinity to the sonic line $M_\infty = 1$ (Fig. 18). The flow in the region of

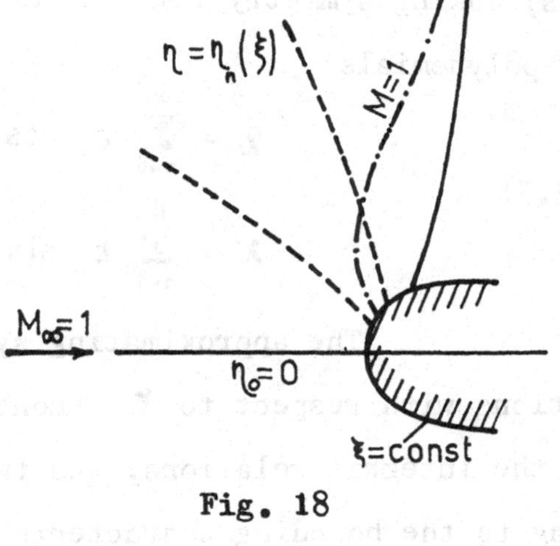


Fig. 18

influence is potential and it is described as before by the system of equations (2.2.1), but in which $H_I$ and $H_{II}$ are the Lamé parameters commensurate with the adopted orthogonal coordinates (2.2.5). On the other hand, in the region now under consideration these equations change their character on the sonic line. The boundary conditions of the problem are the conditions of symmetry on the axis $\eta = 0$, the solid surface condition on the body, and the condition of uniform sonic velocity at infinity; in addition, on the bounding characteristic $\eta = \eta_1(\xi)$, the differential compatibility condition must be satisfied.

In solving this problem, the region of influence is subdivided by the curves

$$\eta_n(\xi) = \frac{N-n+1}{N}\,\eta_1(\xi) \qquad (n = 2,3,\ldots,N) \tag{2.2.6}$$

extending between the bounding characteristic and the axis. The integrand functions $\chi$ and $\lambda$ are represented in the integral relations, taking symmetry into account, by the following interpolation polynomials

$$\chi = \sum_{n=0}^{N} a_n \cos n \frac{\pi}{2}\frac{\eta}{\eta_1}\,, \qquad \lambda = \sum_{n=1}^{N} b_n \sin n \frac{\pi}{2}\frac{\eta}{\eta_1}\,. \tag{2.2.7}$$

The approximating system of ordinary differential equations with respect to $\xi$, contains $2N$ equations following from the integral relations, and two differential equations relating to the bounding characteristics. This system permits the determination of $2N+1$ velocity components on the axis, the bounding characteristic and the intermediate lines as well as the characteristic itself $\eta=\eta_1(\xi)$ (Fig. 18). If the integration of the approximating system is conducted from the body outward to infinity, then in the axisymmetric case it becomes necessary to solve a boundary problem of the $(N+1)$ order (the specified parameters are: $N$ values of velocity $v_n$ and coordinate $\eta_1$ of the bounding characteristic on the body). In the two-dimensional case, in view of an additional relationship which applied (the characteristic in the hodograph plane is an epicycloid), the order of the boundary problem decreases by unity.

In the given scheme the approximating system has just one singular point at infinity, through which, if integration is carried outward from the body, it is necessary to pass. The approximation (2.2.6) assures that the singularity at infinity corresponds to uniform sonic flow and permits satisfaction of the solid surface condition on the entire contour of the body Thus, in every approximation the solution correctly satisfies all the boundary conditions expressed over a correctly defined region of influence.

We shall present certain calculated results for the sonic flow of air ($\gamma = 1.4$) about a circle. Pressure distributions are shown in Fig. 19 for the nose part of a circle as computed in three approximations, the pressure p being referred to the stagnation pressure $p_{st}$ . Also plotted on this graph,

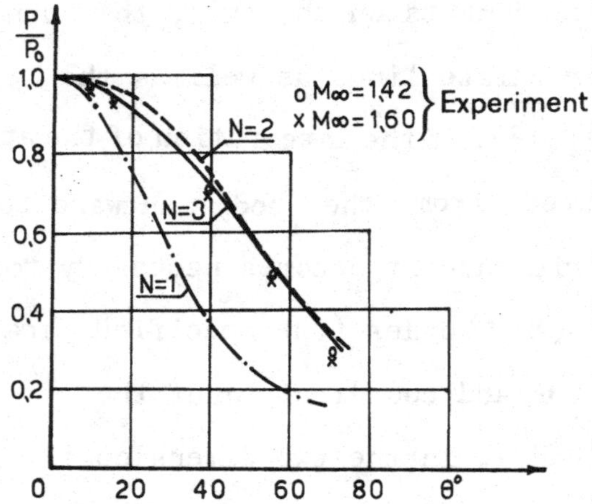


Fig. 19

which refers specifically to $M_\infty = 1$, are the experimental data of D. Holder and A. Chinnick (1954) obtained for a circular contour at $M_\infty = 1.42$ and $M_\infty = 1.60$. It is well known that in the interval of infinite Mach numbers $M_\infty$ close to unity the pressure distributions of a body placed in the flow remain practically unchanged. Consequently, the approximate comparison of theoretical and experimental results herein presented may be considered sufficiently convinving.

## 2.3 Flows with Shock Waves

### 1. Supersonic Flows about Blunt Bodies [13 - 15] [21, 22]

The method of integral relations for the calculation of supersonic flows of a perfect and real gases about the nose regions of two-dimensional and axisymmetric blunt forms was developed by O.M. Belotserkovskii (1957, 1958, 1960, 1962). Here the direct problem of determining the flow about a body of given form is investigated. As an end result, the detached shock wave and the mixed flow in the region of influence between shock wave, body and limiting characteristics are calculated.

The governing system of differential equations is conveniently taken in $s, n$ coordinates, where $s$ is the arc length measured along the body, while $n$ is the normal to the body (the body contour is given by $n = 0$), or in the coordinates $s, \xi = n/\varepsilon(s)$, where $n = \varepsilon(s)$ is the equation of the shock wave. In the $s, n$ system, the form of the body is easily varied, which permits the calculation of flows about bodies with arbitrary contours (smooth or broken) by means of a single program. Furthermore, the approximating system in these coordinates is of simple form, while the resulting data may be directly used for calculating boundary layer or heat transfer. In the $s, \xi$ system the shock wave becomes parallel to the body, which permits the formulation of integral relations along $s$ from the axis of symmetry to the bounding characteristic.

In applying the method of integral relations to the calculation of supersonic flow about blunt bodies two types of problems were examined, the adiabatic flows of a perfect gas with constant specific heats and flows of mixtures of reacting gases in thermodynamic equilibrium and non-equilibrium. The complete system of equations describing such flows consists of the equations of motion (2.3.1), continuity (2.3.2), energy (the adiabatic condition)(2.3.3), equation of state (2.3.4) and kinetic equations (2.3.5) for the $\ell$ components. For the case of non equilibrium dissociation these equations have the form [22]

$$(\vec{V}\nabla)\vec{V} + \frac{\nabla P}{\rho} = 0\,, \qquad (2.3.1)$$

(2.3.2) $$\nabla(\rho\vec{V}) = 0,$$

(2.3.3) $$\frac{dh}{dt} - \frac{1}{\rho}\frac{dP}{dt} = 0,$$

(2.3.4) $$P = R\rho T\sum_{i=1}^{l}\frac{c_i}{m_i},$$

(2.3.5) $$\frac{dc_i}{dt} = \frac{\omega_i}{\rho} \quad (i=1,2,\dots,l)$$

To these should be added expressions for specific enthalpy

$$h = \sum_{i=1}^{l} c_i h_i\,, \quad h_i = h_i' + h_i^0\,, \quad h_i' = \int_0^T c_{pi}\,dT\,.$$

Here $V, P, \rho, T, h$ are velocity, pressure, density, temperature, and enthalpy, respectively; $R$ is the universal gas constant; $\omega_i$ is the mass speed of generation of the component as a result of chemical reaction; $c_i, m_i, c_{pi}$ are the mass concentrations, mole cular weights and specific heats at constant pressure of the ith component, respectively; $h_i^0$ is the enthalpy of formation of the ith component extrapolated to absolute zero, while $d/dt$ is the total derivative with respect to time.

The system (2.3.1) to (2.3.5) may be simplified if all known integrals are incorporated and if some of the partial differentials are written along the stream line in the form of ordinary differential equations. As a result, the governing system will contain fewer differential equations which are, in fact, the only kind which need to be solved by the method of integral relations. Finite relationship and ordinary differential equations of the approximating system are taken along the boundaries of the strips and consequently will be applied in an exact form. In this way, the boundary problems for this approximating system will be in a simpler form and will conform more exactly to the governing system of equations.

In this connection it is expedient to introduce, instead of some of the kinetic equations (2.3.5), equations of material balance and Dalton's equation

$$\sum_{i=1}^{e} c_i = 1 . \qquad (2.3.7)$$

It is also useful to introduce the Bernoulli integral

$$h + V^2/2 = h_{st} , \qquad (2.3.8)$$

where $h_{st}$ is the stagnation enthalpy.

For a perfect gas the adiabatic condition (2.3.3) is conveniently expressed in the form that entropy $S$ is constant along the stream line $\Psi$ = constant:

$$S = S_s(\Psi) , \qquad (2.3.9)$$

where $S$ is the value of entropy directly behind the shock wave. In addition, in this case

$$h = \frac{\gamma}{\gamma-1}\,\frac{p}{\rho} , \qquad S = c_v \ln \frac{p}{\rho^\gamma} + \text{const.} , \qquad (2.3.10)$$

where $c_v$ is the specific heat at constant volume. After this, the variables $p$, $\rho$ , and $T$ are readily eliminated from the governing system, which will consist, as a result, of four equations in the four unknown functions: $u$, $v$ (the components of the velocity $\vec{V}$ along $n$ and $s$ ), $S$, $\Psi$.

For the flow of a mixture of reacting gases in equilibrium the magnitudes of the equilibrium concentrations

$c_i = c_i(p, T)$ are determined as solutions of a non-linear algebraic system of thermodynamic equations resulting from the conditions $\omega_i = 0$. In this case, the governing system (2.3.1) to (2.3.5) need not contain the quantity $c_i$. The enthalpy and equations of state are then written in the following form

(2.3.11) $$h = h(p,T) \qquad \rho = \rho(P,T).$$

For any particular mixture of gases these relationships are obtained from thermodynamic tables which, for convenience in calculation, may easily be approximated. For air, such approximations were constructed, for example, by I.N. Naumova (1961). The adiabatic condition in the equilibrium case is written just as for a perfect gas.

In calculating non-equilibrium flows it is not expedient to introduce the entropy, while the adiabatic condition (if it is to be utilized) and the remaining kinetic equations are conveniently taken along stream lines in the form

(2.3.12) $$dh - \frac{dp}{\rho} = 0, \quad \rho u dc_i = \omega_i dn,$$

in which

(2.3.13) $$dn = \left(1 + \frac{n}{R_b}\right)\frac{u}{v}\, ds,$$

where $R_b$ is the radius of curvature of the body. Let us now examine the various schemes of solution of this problem by the method of integral relations. While these schemes are applicable to flows about blunt bodies at an angle of attack as well, the concrete examples below will be limited to symmetrical flows at zero angle of attack, since the calculations were mainly conducted for this case [22].

Scheme 1. The governing system of equations is written in the $s, n$ coordinate system. In calculating all types of flow in accordance with this scheme, Bernoulli's integral is introduced in place of the equation of motion in the $s$ direction.

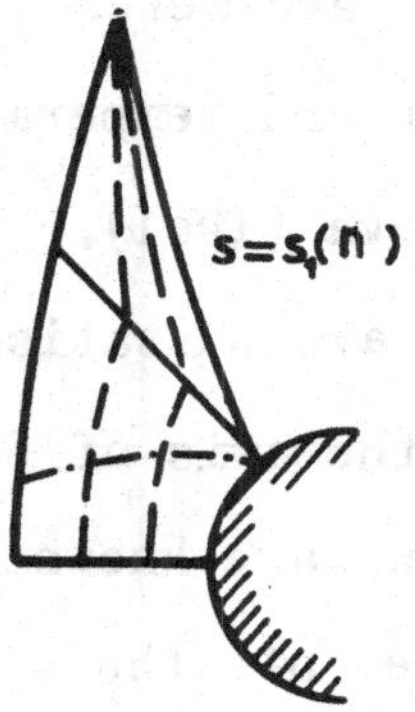


Fig. 20

The region of integration (Fig. 20) is subdivided into strips by lines spaced by constant intervals of $n$, and constructed between the body $n = 0$ (or the bounding characteristic $s = s_1(n)$ and the shock wave $n = \varepsilon(s)$). The integrand functions are replaced by the interpolation polynomials in $n$, which results in an approximating system in $s$.

For example, for equilibrium flows the approximating system in the Nth approximation has, in accordance with scheme 1, the form

$$\frac{d\varepsilon}{ds} = \left(1 + \frac{\varepsilon}{R_b}\right)\tan\left(\sigma - \Theta_b\right), \quad \frac{d\sigma}{ds} = \varphi, \quad \frac{dV_{n1}}{ds} = \frac{dV_{n1}}{d\sigma}\varphi,$$

$$\frac{dp}{ds} = \frac{dp_1}{d\sigma}\varphi, \quad \frac{ds_1}{ds} = \frac{ds_1}{d\sigma}\varphi, \quad \frac{dT_1}{ds} = \frac{dT_1}{d\sigma}\varphi.$$

$$\frac{du_i}{ds} = U_i, \quad \frac{dv_i}{ds} = \frac{E_i}{v_i^2 - a_i^2}, \quad \frac{dp_i}{ds} = P_i, \tag{2.3.14}$$

$$\frac{dT_i}{ds} = \textcircled{H}_i, \quad \frac{d\Psi}{ds} = F_i, \quad s_i(\Psi_i) = s_1(\Psi_i) \qquad (i = 0, 2, 3, \dots, N).$$

Here the values of variables on the shock wave are designated by the subscript 1, those on the body by the subscript 0, and on the intermediate lines $n = n_i(s) = \varepsilon(s)(N-i+1)/N$ $(i = 2, 3, \dots, N)$ by the subscript $i$. The quantities $\sigma$ and $\Theta_b$ denote the angles of inclination of shock wave and body, respectively, to the direction of the incident flow, $V_{n1}$ is the normal component of velocity directly behind the shock wave. The known holomorphic functions $\varphi$, $U_i$, $E_i$, $P_i$, $Ⓗ_i$, $F_i$ comprise the right-hand sides of this system. Enthalpy, density and the sonic velocity $a$ are determined here for the corresponding values of pressure and temperature from approximations of the type given by Naumova (1961).

Boundary conditions on body and shock wave are satisfied exactly in this scheme at each approximation. On the axis of symmetry $s = 0$, all the parameters of the flow are known with the exception of the detachment distance of the shock wave $\varepsilon_0$ and $N-1$ values of velocity $u_i (i = 2, 3, \ldots, N)$. In the supersonic neighborhood of the transition line (at $v_i = a_i$), however, the system has $N$ movable singular points of regular type. It follows from the regularity condition of solution at these singular points, that $E_i = 0$ at $v_i = a_i$ $(i = 0, 2, 3, \ldots, N)$ from which the $n$ arbitrary parameters on $s = 0$ are determined uniquely.

The singular points are characterized by the property that the characteristics there of the governing system of differential equations are tangent to the direction $s$ = constant. This relationship between the characteristics and the singular points is a significant feature of scheme 1.

Approximating systems of analogous form result when calculating the flow of a perfect gas or non-equilibrium flows.

Two variations of scheme I have been found useful. In the first, the region of integration is bounded at the top by the limiting ray $s = s_1 =$ constant, whose location can be made to satisfy the conditions of regularity at all of the singular points.

In the second variation the region of integration is bounded at the top by the bounding characteristics passing between the shock wave and body. This makes it possible to carry out the calculations within the exact region of influence.

Scheme 1 is advantageous for the calculations of flows with high incident Mach numbers $M_\infty$. It yields the most exact determination of the shock wave and of the distribution of gas dynamic parameters on the body. With moderate values of $M_\infty$ sufficient accuracy in the neighborhood of the nose of the body is obtained with $N = 2$, while for hypersonic flows just the first approximation is entirely adequate. Good results are obtained here even if the body is of complex form. Scheme 1 also yields acceptable initial data for further calculation of the supersonic region of flow by the method of characteristics, inasmuch as disturbances in the initial data on the bounding ray or characteristic quickly die out downstream.

The solution by scheme 1 of the supersonic symmetrical flow of a perfect gas about the nose region of a blunt body was developed by O.M. Belotserkovskii for the two-dimensional case (Belotserkovskii, 1957, 1958) and for the axisymmetric case (Belotserkovskii, 1960), polar coordinates being applied here. Scheme 1 was used to calculate the flow about the nose regions of bodies of various contours-smooth, with a discontinuity in curvature, with a sharp corner. The results of calculations for circular cylindars and for ellipsoids of revolution are published, respectively, in Belotserkovskii, (1958 ), and Belotserkovskii, (1961). In addition, calculations using initial data obtained by this scheme of solution were performed by the method of characteristics for the supersonic portion of the flow on blunt bodies (Chushkin, 1960. Belotserkovskii and Chushkin, 1962). In particular, detailed tables of the flow in the neighborhood of blunt cones were computed (Chushkin and Shulishnina, 1961).

The flow of a perfect gas about a blunt body of revolution whose contour possesses a sharp corner was also studied by A.P. Bazzhin (1961), who applied scheme 1 to the nose region in the coordinates $s, \bar{\Psi} = 2\Psi/\rho_\infty v_\infty y_s^2$ (where $\Psi$ is the stream function, $y_s$ is the distance from the axis of the flow to a point on the shock wave), and at the corner of the contour in the coordinates $\Theta, \bar{\Psi}$ (where $\Theta$ is the polar angle).

The application of scheme 1 to the calculation of the flow of a real gas was examined in Belotserkovskii (1962) and Belotserkovskii et al. (1964). In the latter work a numerical solution was obtained (in the second approximation of the method of integral relations) to the problem of symmetric flow about blunt bodies of gas (air, $CO_2$) dissociating in equilibrium. Initial data obtained from Belotserkovskii et al. (1964) were utilized by I.N. Naumova (1963) for calculating supersonic equilibrium flows on blunt bodies by the method of characteristics.

Scheme II. The governing system of equations is written in $s\xi$ coordinates, in which the contours of the body ($\xi = 0$) and the shock wave ($\xi = 1$) are straightened. In the Nth approximation, N-1 intermediate lines equally spaced with respect to $s$ (Fig. 21) are passed between the axis of symmetry $s = 0$ and the bounding characteristic $s = s(\xi)$. Approximating polynomials are constructed in $s$, taking into account the symmetry with respect to $s = 0$. The shock wave is represented by a polynomial of even powers. The Bernoulli integral is introduced in place of the equation of motion in the $s$ direction or in place of the adiabatic condition. The approximating system in this scheme does not have singular points (if approximations assuring regularity of the solution with respect to $s$ are employed). On the other hand, a boundary value problem arises here, with, in the Nth approximation, N+1 arbitrary parameters on the shock wave (the values of $\varepsilon$ on the boundaries of the strips) and N+1 solid surface condition on the body permitting determination of these parameters.

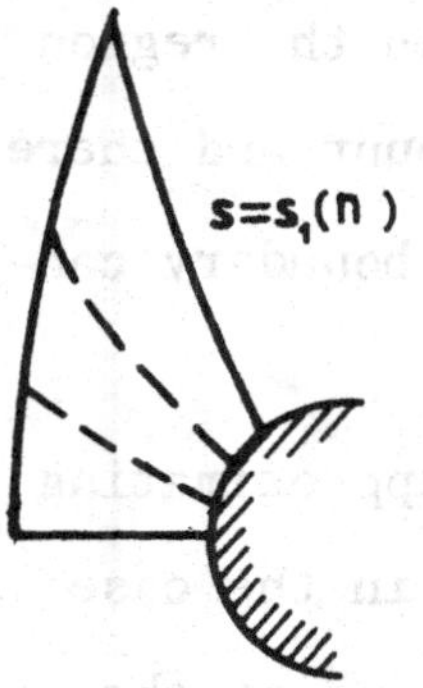


Fig. 21

Thus, in scheme II, the boundary condition on shock wave and body are satisfied only at discrete points (the traces of the boundaries of the strips on shock wave and body ), the number of which grows with the degree of approximation. In this system the region of influence is exactly defined in every approximation, while the approximating system is completely closed, thanks to the differential compatibility relation on unknown bounding characteristic.

Let us note that, generally speaking, in scheme II one cannot bound the region of integration from above by a ray or any other noncharacteristic line, since then the region of influence would not properly be taken into account and there would not be on the upper boundary the additional boundary condition necessary to close the system.

The technique of constructing the approximating system of scheme II is the usual one except that, in the case of axisymmetric flows , it is necessary for functions of the type $F = y(s,\xi) f(u,v,...)$ to carry out separate approximations of the factors $y(s,\xi)$ and $f(u,v,...)$.

Scheme II is advantageous in those cases when the approximating functions change more slowly with $s$ than with $n$. This occurs, for example, with bodies of gradually varying curvature at small values of Mach numbers $M_\infty$ or in non-equilibrium flows, when the distribution of the flow parameters across the shock layer may bear a sharply varying character. In these instances good accuracy may be obtained with scheme II at relatively small $N$. Calculations according to this scheme take less machine time than those in accordance with sceme I.

Scheme II was applied by Chushkin (1957) to the study of the limiting case previously mentioned - the sonic ($M_\infty=1$) flow about bodies, in which the shock wave recedes infinitely far from the body. Calculations were also performed for a weak detached shock wave (the wave is assumed distorted, but the flow behind it potential). Recently O.M.Belotserkovskii and V.K. Dushin (1964) used this scheme to calculate the flow of a gas in non-equilibrium dissociation about blunt bodies in the symmetrical case.

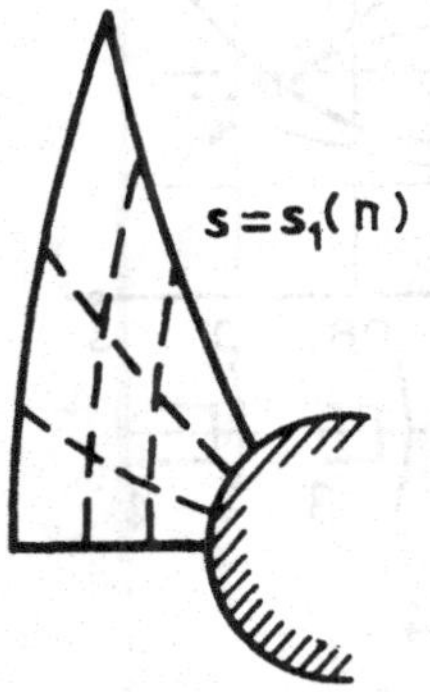

Fig. 22

Scheme III. The region of integration is divided into sub-regions (Fig. 22) . The functions are approximated in two directions and the problem reduces to the solution of a system of transcendental equations. This scheme may prove useful in certain cases. On the other hand, it should be born in mind that the results obtained with its use are less accurate than those obtained by the two previous schemes.

## 2. Calculating Mixed Flows

We shall now present a series of curves illustrating the computed supersonic symmetrical flows about blunt bodies, as obtained by O.M. Belotserkovskii and his colleagues.

First in order will be results referring to the flow of a perfect gas (Figs. 23-26); these are followed by the flows of equilibrium and non-equilibrium dissociating gas [22]. Linear dimensions here are referred to the radius of the body.

In Fig. 23 there is shown the pressure distribution $\bar{p}_b$, on a body referred to the pressure at the

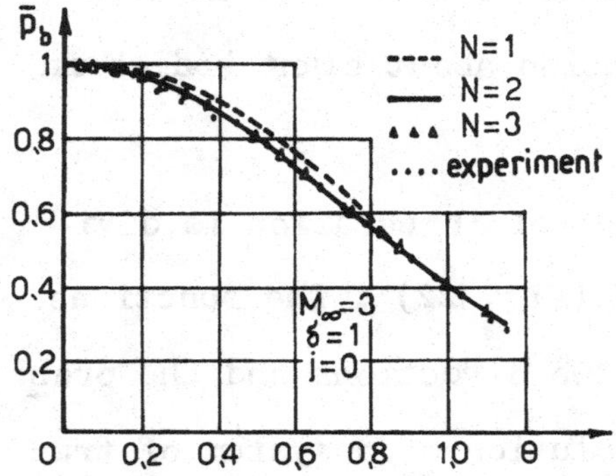


Fig. 23

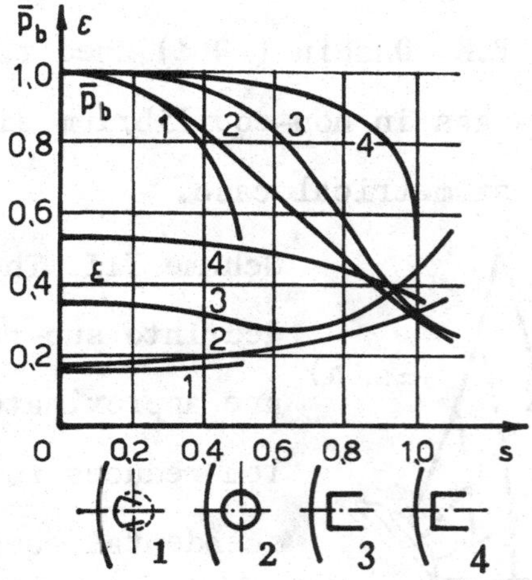


Fig. 24

stagnation point $p_{st}$, for the case of flow about a circle of a perfect gas ($\gamma = 1.4$) at Mach number $M_\infty$=3. Here results are shown in three approximations: N=1,2,3 (scheme 1) and the experimental points of G.M. Ryabinkov. This graph characterizes the rapid convergence of the method of integral relations. It should be borne in mind that the case under consideration (plane body and relatively small $M_\infty$) is a difficult one from the point of view of the method, bacause of the great thickness of the shock layer. The distribution of pressure $\bar{p}_b$, the variation in the distance $\varepsilon$ to the shock wave as a function of $s$ for axisymmetric bodies of various forms (sphere with a break in the contour, smooth sphere, cylinder with blunt leading face

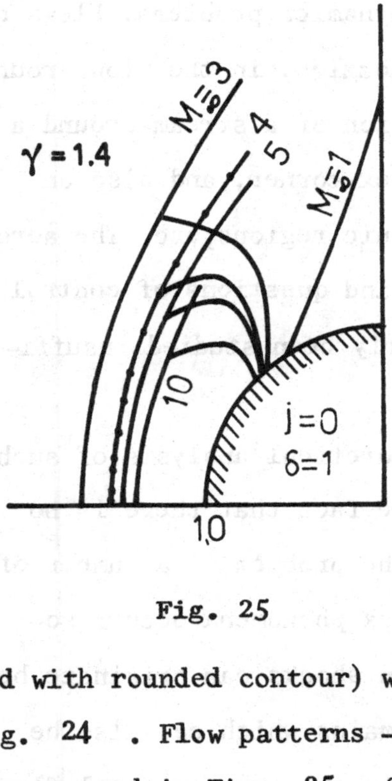


Fig. 25

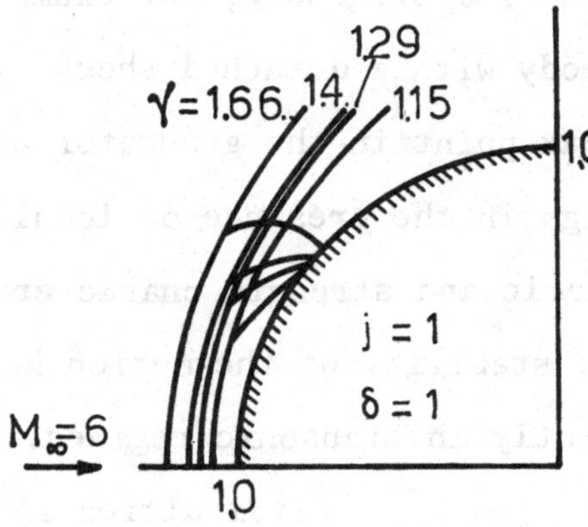


Fig. 26

and with rounded contour) with $M_\infty = 4$ and $\gamma = 1.4$ are shown in Fig. 24 . Flow patterns – shock waves and sonic lines – are constructed in Fig. 25 for a circle with $\gamma = 1.4$ in the range of Mach numbers $M_\infty = 1 \div 10$ (here also are shown the experimental points of Kim (1956) for $M_\infty = 4$ ) and for a sphere in Fig. 26 at $M_\infty = 6$ and various values of the adiabatic index $\gamma$.

## 2.4 Transonic Flows Behind a Detached Shock [22], [29], [31]

A great deal of attention has recently been paid to the investigation of transonic gasdynamics problems. Flows of this kind originate, for example, in nozzles, in the flow around a body with a detached shock, in rotation of a stream around a break point in the generator of a convex corner, and also on wings in the presence of local supersonic regions,etc. The aerodynamic and strength characteristics, and questions of control and stability of the motion have clearly been studied insufficiently in transonic regimes.

Difficulties in the theoretical analysis of such problems arise primarily because of the fact that there is no complete mathematical formulation of the problem in a number of cases. In mixed gas flow domains complex phenomena occur: so--called secondary "hanging" compression shocks can originate behind the bow wave; local supersonic domains which can also be closed off by compression shocks form in the post-critical flow around bodies, etc. The mechanism for the generation of such phenomena has not been studied sufficiently; the question of the kind of minimal region of influence and of its reorganization with the change in flow parameters also remains open etc.

Furthermore, it should be noted that analytical and numerical methods have not been developed, in practice, for vortical plane and spatial transonic flow. The classical analytical methods developed for plane potential flows, and based predominantly on utilization of the hodograph plane, and a different kind of simplification and linearization, do not pass here, as a rule. The numerical solution of the mentioned problems causes many difficulties, and requires the construction of special schemes for the integration of equations of elliptic-hyperbolic type.

Some transonic gasdynamics problems, referring to the theory of plane and spatial vortical perfect gas flows which have been treated mainly by the authors, or under their supervision, are considered herein.

Results of analytical investigations of a number of exact properties of the solutions of problems of the flow around bodies with a detached shock are presented in the first part of the paper. The results obtained are presented without proof in the form of theorems (there are proofs in the papers cited, or will soon be published). Numerical schemes are considered in the second part, and results are presented of computations in the solution of the complete problem of supersonic flow around blunt bodies with a break in the contour generator, and also at low supersonic velocities. The boundaries for the existence of distinct kinds of minimal domains of influence of mixed flow at the bluntness are also studied here.

Only numerical schemes utilizing high-speed

computers and carefully conducted experiments permit quantitative data and a complete flow picture to be obtained in such complex problems. An attempt is made here to analyze and discuss some transonic flow properties (the solutions in the neighborhood of the sonic lines, the formation of secondary compression shocks, the shape of the minimal domain of influence, etc.) from the aspect of analytical and numerical solutions. It can be mentioned that precisely by numerical means are secondary compression shocks, non-monotone sonic lines, etc. successfully detected and constructed. At the same time, the conditions for spoiling the continuous solution, and the regularities associated with different kinds of domains of influence, have been detected analytically. In examining the flow around a corner, the effective construction of numerical schemes turned out to be possible only when utilizing asymptotic methods of solution of the differential equations, etc. Therefore, the combination of both directions turns out to be quite fruitful. Research in these areas continues.

An analysis of transonic problems with the utilization of numerical methods is of indubitable interest. In the majority of cases, a complete solution of the non-linear problem is successfully obtained here, and the required flow characteristics are determined.

If there is a break in the generator in the M-domain the flow around a blunt body, and the flow pattern is altered so that the local speed of sound is achieved at the break-point, then the analysis of such a flow is complicated substantially by the existence of the singularity. The zone of stream rotation around the corner will be in the mixed transonic flow region, where the rotation itself is accompained by an abrupt change in velocity in both magnitude and direction. Moreover, a secondary compression shock, affecting the whole flow pattern substantially, can originate in the supersonic zone of the flow around the lateral body surface. It should also be noted that the results of computing the flow in the domain of influence of the bluntness are later the initial data for the calculation in the supersonic zone, where construction of the solution in the transonic domain for a body with a break in the contour must be made with special care since even slight inaccuracies in the calculations will not permit continuation of the computation into the supersonic domain.

A study of the properties for low supersonic flows around blunt bodies is of no less interest. As the free stream Mach number decreases, the domain of mixed flow influence increases. In examining such a problem it is necessary to take account of the transonic nature of the flow in the zone between the sonic line and the boundary characteristic separating the M-domain. Perturbations in the transonic domain affect the shape of the sonic line, and therefore, the whole flow in the mixed zone; the solution of the boundary value problem becomes more and more responsive to a change in initial data, round- off errors increase, and an instability in the computation is manifest. All this demands the construction of special numerical schemes; it is hence important to represent the boundaries of the existence of distinct kinds of minimal domains of influence of the bluntness.

A number of papers has been devoted to the study of these questions. We consider here the results of investigations of the direct two-dimensional (plane or axisymmetric) problem obtained by using the method of integral relations.

## 1. Supersonic Flow around Blunt Bodies with a Break in the Generator of the Contour [22, 29].

Scheme I or scheme II of the Dorodnitsyn-Belotserkovskii method of integral relations, whose computational mashes are shown in Fig. 20 and 21.

The integral relations method is applied in the first scheme in both the domain up to the break, and in the supersonic rotation around it, where the initial equations are written in a polar coordinate system with center at the angular point. The system of ordinary differential equations is hence integrated numerically along the shock layer from the axis of symmetry. In the neighborhood of the break, where the Prandtl-Mayer solution holds, a differential relation is used which is the compatibility condition along the second family characteristic.

In the second case (Fig. 20), the domain the bluntness is successfully isolated exactly because of the construction of the boundary characteristic, and this substantially increases the accuracy of the calculation in each approximation. Here the asymptotic Vaglio-Laurin-Shugaev solution, reduced to a form convenient for calculations is utilized in the neighborhood of the break (the domain G).

The solution which describes the plane and axisymmetric transonic flow of a perfect gas in the neighborhood G of an angular point has the form of a power series in the distance $n_0$ from the body surface ($n_0 = 0$) with coefficients dependent on the corresponding self-similar variable $\zeta$ [22, 29]

$$u = 1 + (\gamma+1)^{-1/3} \sum_{i=0}^{\infty} u_i(\zeta) n_0^{(2+i)/4}, \quad V = \sum_{i=0}^{\infty} v_i(\zeta) n_0^{(3+i)/4},$$

$$\zeta = (\gamma+1)^{-1/3} s_0 n_0^{-5/4}, \quad n_0 = n/r_0^*, \quad s_0 = (s - s^*)/r_0^*, \tag{2.4.1}$$

where $U$, $V$ are velocity components (referred to the critical speed of sound) along the tangent and normal to the body surface in the subsonic neighborhood of the angular point; quantities at the sonic point on the body are denoted with an asterisk.

The main term of this expansion describes two--dimensional transonic potential flow in the neighborhood of the angular point of the profile (primes denote derivatives with respect to $\zeta$)

(2.4.2) $$u_0 = g', \quad v_0 = (7g - 5\zeta g')/4,$$

where

$$g'' = B_1(21g - 25\zeta g')/16, \quad B_1^{-1} = g' - (25\zeta^2)/16.$$

It is interesting to note that the function corresponding to the potential of this flow, is expressed in the parametric form [29]

$$g = (25/42)\, 5^{-1/8}(5z^2 + 5z - 4)(1-z)^{-7/8}(1+3z/5)^{-9/8}C^{-3},$$

(2.4.3)

$$\zeta = -2\cdot 5^{-3/8} z(1-z)^{-5/8}(1+3z/5)^{-3/8}C^{-1}, \quad -5/3 < z < 1.$$

If the scale factor is $C = 1$, then $f^1(1)=0$, where $g \sim (125/56)2^{-1/5}(-\zeta)^{7/5}$ for $\xi \to -\infty$; $g \sim \zeta^5/3 - 675/96\, 10^{-1/3}\xi^{1/3}$ for $\xi \to +\infty$.

Members of the expansion (2.4.1) of higher order of smallness, which take account of the vortical and axisymmetric nature of the flow ($u_i, v_i, i = 1,2,\ldots$) are found by solving linear inhomogeneous ordinary differential equations. The functions $u_i, v_i$ should here satisfy the boundary conditions in both the subsonic domain (zero normal velocity component on the body surface), and in the supersonic domain where the solution describes the Prandtl-Mayer type flow.

An analytic solution of this system was found in [29] by F.V. Shugaev.

Let us represent the quantities $u_i, v_i$ as the sum

of the particular solution $u_i^{(1)}, v_i^{(1)}$ of the inhomogeneous system, and the general solution $u_i^{(2)}, v_i^{(2)}$ of the homogeneous system.

If $\Phi_i(\xi)$ is introduced, so that the solution $u_i^{(2)}$, $v_i^{(2)}$ of the corresponding homogeneous solution would be written as

$$u_i^{(2)} = \frac{d\Phi_i}{d\zeta}, \quad v_i^{(2)} = \frac{7+i}{4}\Phi_i - \frac{5}{4}\zeta\frac{d\Phi_i}{d\zeta}, \qquad (2.4.4)$$

then a second order equation is obtained to determine $\Phi_i$

$$(2.4.5) \qquad (16g' - 25\zeta^2)\Phi_i'' + \left[16g'' + 5(5-2i)\zeta\right]\Phi_i' - (7+i)(3-i)\Phi_i = 0.$$

Utilizing the parametric representation (2.4.3) of the function $g(\zeta)$ and making the change of variable

$$t_i = \Phi_i(1-z)^{(7+i)/8}\left(1+\frac{3}{5}z\right)^{3(7+i)/8}, \qquad (2.4.6)$$
$$\xi = 1/4\left[2 - \sqrt{(3/2\,(1-z))}\right],$$

we obtain the hypergeometric equation

$$6\xi(1-\xi)t_i'' + (9+2i)(1-2\xi)t_i' - 2(7+i)(9+i)t_i = 0. \qquad (2.4.7)$$

Its solution is the Jacobi polynomials

$$t_i = \left[\xi(1-\xi)\right]^{(5/2+i/3)}\frac{d^{(7+i)}}{d\xi^{(7+i)}}\left\{\left[\xi(1-\xi)\right]^{(2i/3+9/2)}\right\}, \qquad (2.4.8)$$

which permit expression of the quantities $u_i, v_i$ in final form.

Such a representation turns out to be quite convenient for calculations in the domain G since direct integration of the original system would evoke definite difficulties. Computations have shown that it is sufficient to use the first two-three terms of the solution (2.4.1)-(2.4.8) to construct the flow in the neighborhood of an angular point; the influence of the remaining terms is negligible.

In connection with the fact that the boundary characteristic of the domain of influence of the bluntness enters the break at the supersonic point, the expansion (2.4.1), which is valid only near the sonic line in the supersonic domain, turns out to be inapplicable here. Hence, it is necessary to continue the solution into the purely supersonic domain, which has indeed been found as a correction in powers of $r_1$ of the type $\sum f r_1^{\alpha_i}$ to the fundamental Prandtl-Mayer solution.

The customary algorithm of scheme II is applied outside the domain G : Systems of ordinary differential equations are integrated across the shock layer between the wave and the body, where additional "gluing" conditions for both solutions on the boundary of this domain make the problem single-valued.

On the whole, computational algorithm permitting a solution to be found with a high degree of accuracy have successfully been obtained. The solution outside the zone of influence of the bluntness is constructed by the usual method of characteristics. Let us present some results of computations obtained by A. Bulekbaev, V.F. Ivanov, E.S. Sedova and F.V. Shugaev [22, 29] .

Constructed in Fig. 27a are the shocks, sonic lines and boundary characteristics for perfect gas flow ($\gamma = 1.4$) around a spherical segment with a sonic break with a $\chi = 30^\circ$ ($M_\infty = 4$) half-angle, and for $\chi = 33^\circ$ and $43^\circ 30'$ ($M_\infty = 10$) in Fig. 27b.

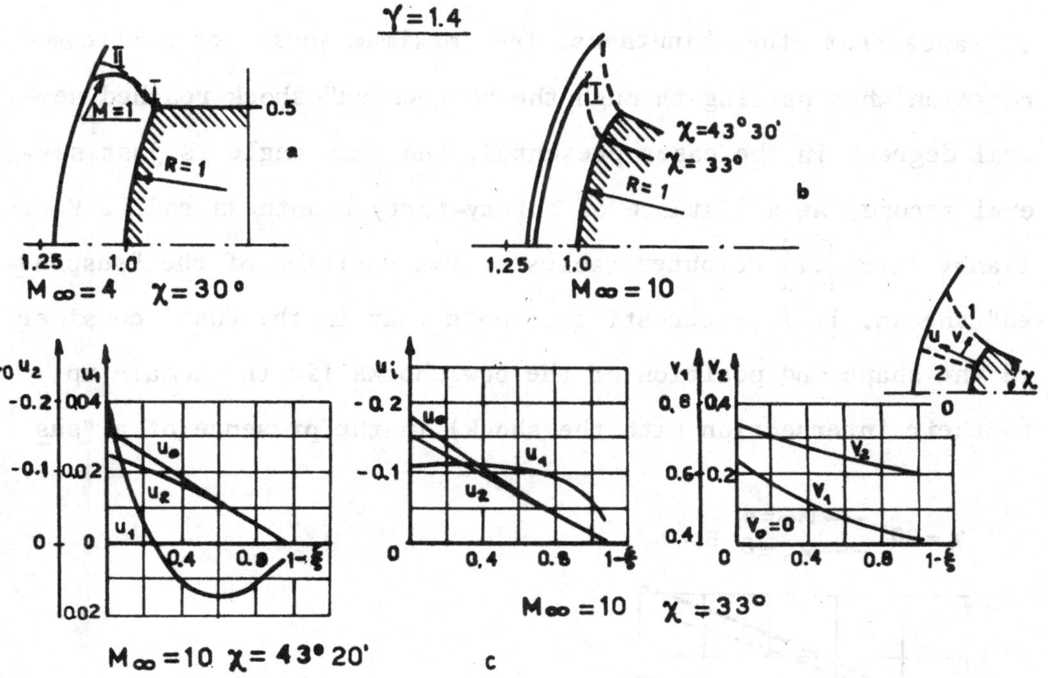


Fig. 27

The distribution of the velocity components (along $n, s$ respectively) on the axis of symmetry ($i = 0$), the boundary characteristic ($i = 1$), and the intermediate line ($i = 2$) is given in Fig. 27c. It is seen that the behavior of $u_1$ on the boundary characteristic depends strongly on the half-angle of the segment.

Presented in Fig. 28 are the shocks and "suspended" shocks (dashes) which originate in the supersonic zone in the flow around cones with spherical bluntness and a sonic break ($\chi = 30^\circ$, $\gamma = 1.4$). Cones with different half-angles ($\omega = -5^\circ, 0, 10^\circ$) and a $M_\infty = 4$ free-stream Mach number are examined.

The intensity of the "suspended" compression shock first increases rapidly, and then gradually drops with distance from the bluntness. The maximum angle of a stream rotation when passing through the "suspended" shock reached several degrees in the cases presented, and this angle is just several seconds at a distance of thirty-forty bluntness radii. V.F. Ivanov first [22] computed tables of the position of the "suspended" shocks. It is interesting to note that in the cases considered the shape and position of the bow shocks (in the domain up to their intersection with the shock) in the presence of a "sus

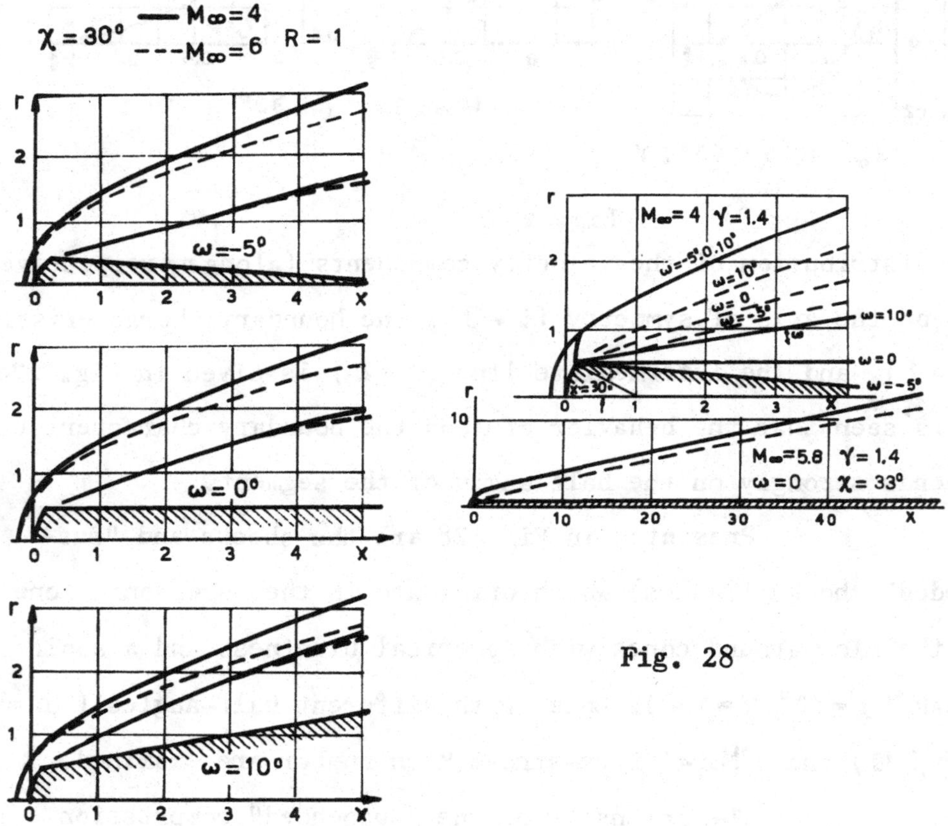


Fig. 28

pended" compression shock will coincide (for the same values of the free-stream Mach number) for different cone half-angles in the $-10° < \omega < 10°$ range of variation. The appearance of the "suspended" compression shock is explained by overexpansion of the stream when the flow turns around the angular point, and its subsequent compression and deceleration at the side surface of the body. Questions associated with the formation of supspended compression shocks in the plane case were discussed in [22] .

## 2 Flow around Blunt Bodies at Low Supersonic Velocities [22]

As has already been remarked, the construction of special numerical schemes would be required for the computation of mixed gas flows at low supersonic velocities. In this case, schemes III of the method of integral relations turned out to be most effective, wherein a representation of functions in two directions is used, and the original equations are approximated by a nonlinear system of algebraic equations in a curvilinear computational mesh (Fig. 29).

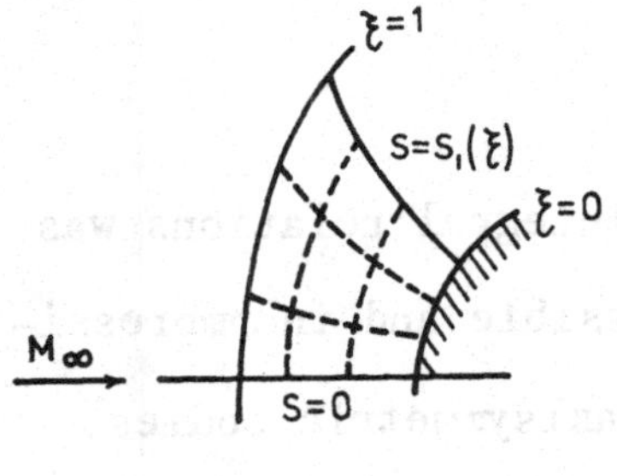


Fig. 29

Utilizing these schemes, F.D. Popov [22] carried out calculations up to free stream Mach number of $M_\infty = 1.05$. (Fig. 30). The complete system of gas dynamics equations was considered here taking vorticity into account.

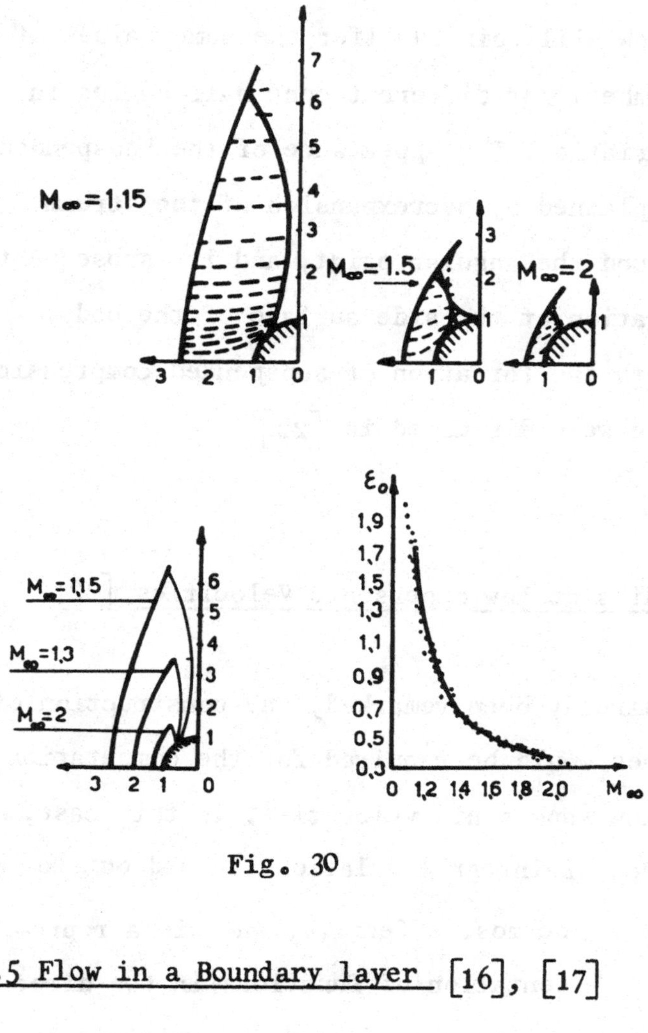


Fig. 30

## 2.5 Flow in a Boundary layer [16], [17]

The generalized method of integral relations was applied to the calculation of both compressible and incompressible laminar boundary layers on plane and axisymmetric bodies, with radiation and heat conduction taken into account.

Let us examine the plane case, bearing in mind that by eans of a well-known transformation the boundary layer equations for the axisymmetric case may be reduced to the form of the equations for the plane case. The governing system of equations for a compressible laminar plane boundary layer may be transformed with the aid of the Dorodnitsyn variables

$$\xi = \int_0^x U \frac{p}{p_{st}} dx , \quad \eta = \frac{U}{\sqrt{\nu_{st}}} \int_0^y \frac{\rho}{\rho_{st}} dy \tag{2.5.1}$$

to the form

$$u \frac{\partial u}{\partial \xi} + w \frac{\partial u}{\partial \eta} = \frac{\dot{U}}{U} \frac{h-u^2}{\bar{T}} + \frac{\partial}{\partial \eta}\left(b \frac{\partial u}{\partial \eta}\right), \tag{2.5.2}$$

$$\frac{\partial u}{\partial \xi} + \frac{\partial w}{\partial \eta} = 0, \tag{2.5.3}$$

$$u \frac{\partial h}{\partial \xi} + w \frac{\partial h}{\partial \eta} = \frac{1}{P_r}\left(b \frac{\partial h}{\partial \eta}\right) - \bar{U}^2\left(\frac{1}{P_r} - 1\right) \frac{\partial}{\partial \eta}\left(b \frac{\partial u}{\partial \eta}\right)^2 \tag{2.5.4}$$

In this system the following notation has been adopted

$$w = \frac{\tilde{U}}{U\sqrt{\nu_{st}}} + \frac{\dot{U}}{U} \eta u , \quad h = u^2 \bar{U}^2 + \bar{T} ,$$

$$\dot{U} = \frac{dU}{d\xi} , \quad \bar{U}^2 = \frac{U^2}{2 \mathfrak{J} C_p T_{st}} ,$$

$$\bar{T} = \frac{T}{T_{st}} , \quad b = b(\bar{T}) = \frac{\nu \rho}{\bar{T} \nu_{st} + \rho_{st}}$$

$U$ – velocity at the outer edge of the boundary layer,

$u$ – tangential component of velocity referred to $U$,

$\tilde{U}$ – a quantity related to the normal component of velocity,

$p, \rho, T, h$ – pressure, density, temperature and enthalpy, respectively,

$\nu$ – coefficient of kinematic viscosity,

$P_r$ – Prandtl number,

$J$ – the mechanical equivalent of heat,

$C_p$ – specific heat at constant pressure.

The subscripts $st$ and $e$ refer to a stagnation point and the outer edge of the boundary layer, respectively.

The system (2.5.2) and (2.5.4) has boundary conditions of the following type

$$u = w = 0 \quad \text{for } \eta = 0$$

(2.5.5)

$$u \to 1, \quad h \to 1 \quad \text{for } \eta \to \infty.$$

To these must be added the temperature condition on the wall (that is, at $\eta = 0$) which depends upon the specific problem and comprises a given relationship involving the temperature, its gradient and the shear at the wall.

Initial conditions must also be applied to the system (2.5.2) to (2.5.4). If the boundary layer begins at $x = 0$ (corresponding to the apex of a body or the stagnation point on a blunt shape) the initial conditions are given by the velocity and temperature profiles of the corresponding self-similar solution for the boundary layer.

The numerical solution of the parabolic system of boundary layer equations (2.5.2) to (2.5.4) will carried out by the generalized method of integral relations. In formulating the integral relations eq.(2.5.3) is multiplied by a certain function $f(u)$

which will be defined below and eq.(2.5.2) by $f'(u)$ , after which the two results are added. The resulting combination is integrated with respect to $\eta$ across the boundary layer and after a change to the variable $u\,(0 \leq u \leq 1)$ a first integral relation is obtained

$$\frac{d}{d\xi}\int_0^1 Ⓗ\, u f(u)\, du = \frac{\dot{U}}{U\bar{T}_e}\left[\int_0^1 Ⓗ\,(1-u^2)\, f'(u)\, du - \right. \tag{2.5.6}$$

$$\left. - \int_0^1 Ⓗ\,(1-h)\, f'(u)\, du - \frac{b_0 f'(0)}{Ⓗ_0} - \int_0^1 \frac{b}{Ⓗ} f''(u)\, du \right.\,.$$

Here $Ⓗ = 1/\frac{\partial u}{\partial \eta}$ is a quantity inversely proporti onal to the coefficient of wall shear; the index "o" designates the values assumed by the relevant variables on the surface of the body. Similarly, multiplying eq. (2.5.2) by $(1-Ⓗ)\, f'(u)$ , eq. (2.5.3) by $(1-Ⓗ)\, f(u)$ , and eq. (2.5.4) by $f(u)$, adding and integrating with respect to $\eta$ across the layer, one obtains the second integral relation (not given here). These two integral relations contain two unknown functions Ⓗ and $\omega = Ⓗ(1-h)$. Once these functions have been obtained, the other pertinent characteristics of the boundary layer are readily derived.

For the particular case of an incompressible liquid we have

$$h = 1\,, \quad \bar{T}_e = 1\,, \quad b = 1\,, \quad \xi = \int_0^x U dx\,, \quad \eta = \frac{Uy}{\sqrt{\nu_{st}}}$$

Here there is but one unknown function, Ⓗ , and for its determination there is the single integral relation (2.5.6), which in this case is written

$$\frac{d}{d\xi}\int_0^1 Ⓗ\, u f(u)\,du = \frac{\dot U}{U}\int_0^1 Ⓗ\,(1-u^2) f'(u)\,du - \tag{2.5.7}$$
$$- \frac{f'(0)}{Ⓗ_0} - \int_0^1 \frac{1}{Ⓗ} f''(u)\,du .$$

The function $f(u)$ is required to vanish sufficiently rapidly to ensure the convergence of the integrals. It is known that as $u \to 1$ $Ⓗ = 0(1/(1-u))$, and hence for an incompressible fluid the system of functions $\{f_n(u)\}$ for the $Nth$ approximation can be the system of powers $\{(1-u)^n\}$, $n = 0,1,\dots,N$. Subdivision of the region $0 \leq u \leq 1$ into strips is accomplished by the lines $u_n =$ = constant, while the integrand functions are represented by the expressions

$$Ⓗ = \frac{1}{1-u}\sum_{n=0}^{N-1} A_n u^n , \quad \frac{1}{Ⓗ} = (1-u)\sum_{n=0}^{N-1} B_n u^n \tag{2.5.8}$$

The set of integral relations formulated for each function $f_n(u)$ yields for the determination of the quantities $Ⓗ_n$, after substitution of the approximations (2.5.8), a system of ordinary differential equations in $\xi$, for which a Cauchy problem must be solved.

With a compressible boundary layer $2N-1$ integral relations are formulated; the same system of functions $\{f_n(u)\}$ is incorporated therein, but in addition to the approximations (2.5.8), in this case approximations are also made of other integrand functions.

In particular, the function $\omega$ is represented by

$$\omega = \sum_{n=0}^{N-1} C_n u^n . \tag{2.5.9}$$

The approximating system consists in this case of ordinary differential equations for the functions Ⓗ and $\omega_n$.

The above method was developed by A.A.Dorodnitsyn (1960 , 1962) for the calculation of an incompressible laminar boundary layers. Here approximating systems were constructed for $N = 1,2,3,4$, and calculations were carried out giving boundary layer characterics at a sufficient number of points (in particular, velocity profiles up to the separation point). Y.N. Pavlovskii (1962, 1964) applied the method of integral relations to the calculations of compressible boundary layers in very general cases. He developed a scheme of solution for boundary layers on plane and axisymmetric pointed and blunted bodies with heat conduction and radiation taken into account at an arbitrary constant Prandtl number Pr and arbitrary dependence of viscosity upon temperature. [17]

It should be noted that the method of integral relations can, without particular difficulties, be extended also to boundary layers in which the specific heat at constant pressure $C_p$ and the Prandtl number Pr are not constant but are given functions of temperature.

We shall quote here the results of Y.N. Pavlovskii (1964) who calculated the compressible plane boundary layer on a plate with a rounded leading edge at Mach number $M_\infty=4$ and Prandtl number $Pr = 0{,}75$. The stagnation temperature was assumed to be $T_{st}= 1000°C$. Sutherland's relationship between viscosity and temperature was assumed and the flow parameters of an ideal gas near the given body were taken from Belotserkovskii [14]

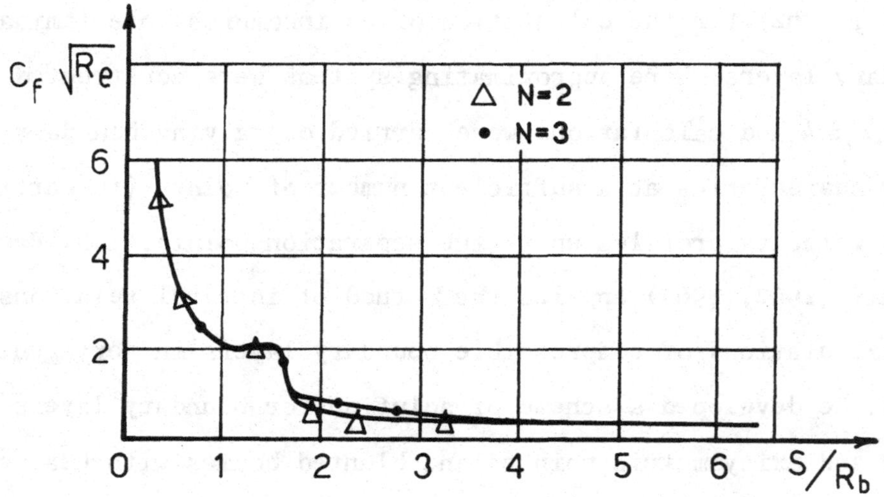


Fig. 31

and Chushkin [23] . The author performed calculations for a non-conducting body, both with and without radiation. The numerical solution was determined in the approximations N=2,3,4,5, the results of these computations give a complete illustration of the convergence of the method.

For the non-radiating case, the variation of $C_f\sqrt{R_e}$

on the body surface with coordinate $\bar{s} = {}^{s}/_{R_b}$, is shown in Fig. 31. Here $C_f$ is the skin friction coefficient referred to $\frac{1}{2} \varrho_e u^2$, $R_e = V_{max} {}^{R_b}/_{\nu_{st}}$ is the Reynolds number, $V_{max}$ is the maximum adiabatic velocity, $R_b$ is the radius of the nose part of the body, the coordinate $\bar{s}$ is measured from the leading edge of the body, $\bar{s} = \frac{\pi}{2}$ represents the point at the end of the nose portion. The functional relationship in the fourth and fifth approxomations are not shown in Fig. 31, inasmuch as they cannot be distinguished on the graph from the data for $N = 3$.

In Fig. 32 there is shown the variation in wall temperature $T_w$ as function of the coordinate $\bar{s}$. Results are present

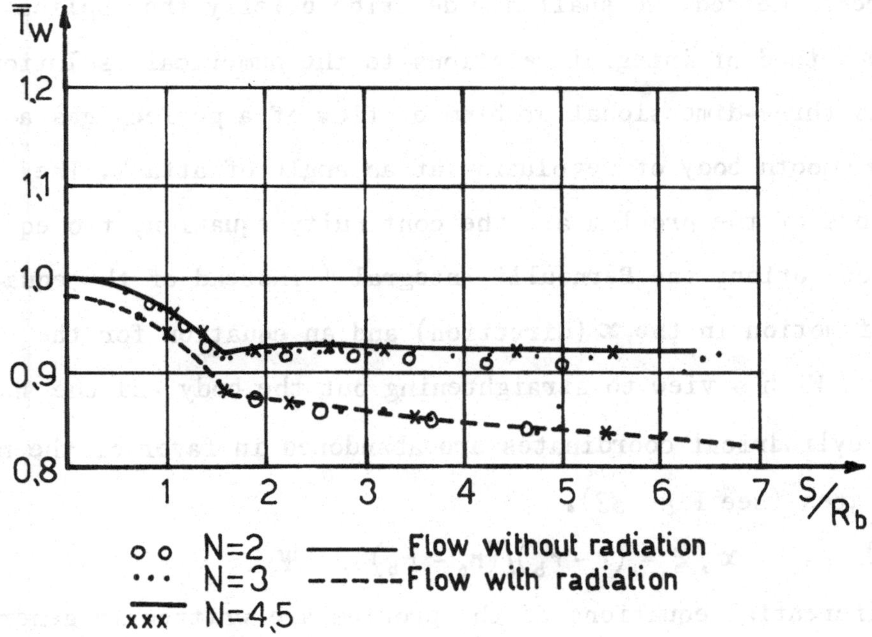


Fig. 32

here for $N=2,3,4,5$ for flows without radiation (the solid lines) and with radiation (dotted lines); for the radiating case the altitude of flight was assumed to 7,700 metres, the radius of the nose $R_b = 1$ metres, the emissivity of the body = 0.6.

## 2.6 Supersonic Flow about a Body of Revolution at an Angle of Attack. [20], [22].

Once again we return to the numerical solution of the purely supersonic region of three-dimensional flows about bodies. In [2] this type of problem was solved by the finite difference method. We shall now describe briefly the application of the method of integral relations to the numerical solution of this three-dimensional problem of flow of a perfect gas about a smooth body of revolution at an angle of attack. The equations of the problem are the continuity equation, two equations of motion, the Bernoulli integral ( instead of the equation of motion in the $x$ (direction) and an equation for the entropy. With a view to straightening out the body and the shock wave, cylindrical coordinates are abandoned in favor of the new coordinates (see Fig. 33).

(2.6.1) $$x\,,\ \xi = (r - r_b)/(r_s - r_b)\,,\quad \Psi\,.$$

The differential equations of the problem are written in general divergence form as follows

$$\frac{\partial N}{\partial x} + \frac{\partial p}{\partial \xi} + \frac{\partial Q}{\partial \Psi} = F \tag{2.6.2}$$

where all of the functions depend on $x$, $\xi$ and $\Psi$ .

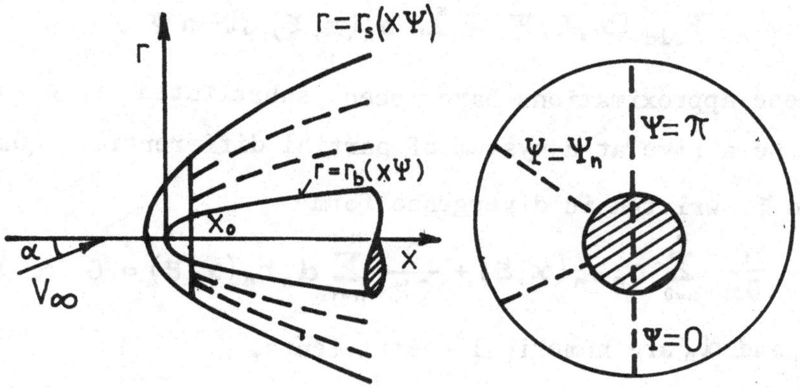


Fig. 33

The usual boundary conditions are applied to these equations– the surface condition on the body, the known conditions on the shock and initial conditions at $x = x_0$. Let us consider a body of revolution $r = r_b(x)$, the flow field for which is to be calculated in a region defined by the inequalities

$$x > x_0 \ , \quad 0 \leq \xi \leq 1 \ , \quad 0 \leq \Psi \leq \pi \ . \tag{2.6.3}$$

The solution of the flow about a body of arbitrary cross-section involves additional technical complications.

Let us draw N-1 equally spaced meridian planes $\Psi = \Psi_n$ = constant across the region of flow. Then even and odd functions of $\Psi$ in the equations of the form (2.6.2) are represented in terms of their values at these meridian planes by the trigonometric expressions.

$$(2.6.4)\qquad \begin{aligned} \mathcal{F}_{even}(x,\xi,\Psi) &= \sum_{n=0}^{N} a_n(x,\xi)\cos n\Psi , \\ \mathcal{F}_{odd}(x,\xi,\Psi) &= \sum_{n=1}^{N-1} b_n(x,\xi)\sin n\Psi . \end{aligned}$$

Once these approximations have been substituted into eqs. (2.6.1) we arrive at a system of partial differential equations in $x$ and $\xi$, written in divergence form

$$(2.6.5)\qquad \frac{\partial}{\partial x}\sum_{n=0}^{N} c_n a_n(x,\xi) + \frac{\partial}{\partial \xi}\sum_{n=1}^{N-1} d_n b_n(x,\xi) = G(x,\xi)$$

where $c_n$ and $d_n$ are numerical coefficients.

In this fashion the governing system of three-dimensional equations is reduced to a two-dimensional system relating functions expressed on separate meridian planes. To the latter system is applied the method of integral relations developed for problems in two dimensions and in which approximations are carried out in the $\xi$ direction by means of interpolation polynomials. One obtains finally an approximating system of ordinary differential equations in $x$, relating a series of functions on lines formed by the traces of shock wave and body surface on $n+1$ meridian planes and on intermediate lines lying in the same meridian planes. The Cauchy problem is solved for this system.

For the special case of supersonic flow about a

body of revolution at zero angle of attack, the corresponding approximating system is derived from the general approximating system if one adopts therein the condition that all the functions are independent of $\Psi$ .

The solution of the system of the two-dimensional partial differential equations (2.6.5) of hyperbolic type can also be carried out by the numerical method of characteristics, which in this case is carried out on individual interrelated meridian planes. It is necessary in this case to use a method of characteristics, in which the solution is constructed in layers bounded by planes $x$ = constant and in which the characteristics on each new layer pass through selected points with values $\xi$ = costant. Such a scheme for use of the method of characteristics for calculating two-dimensional isentropic flows of a perfect gas is described by Katskova and Chushkin (1964) [20].

It should be noted that the three-dimensional method described above may also be used to solve the two-dimensional problem of supersonic flow about a cone at an angle of attack, the solution being determined as a result of establishing self-similarity with respect to $x$.

It is also worth noting that the given scheme for the method of integral relations is directly applicable to the calculation of the subsonic region near the nose part of a blunt body in the three-dimensional case, when it is evidently expedient to introduce the coordinates $s, n, \Psi$ . On the other hand, this leads to difficulties mentioned earlier, related to the necessity of solving a high-order boundary value problem for a system of ordinary differential equations. This work was done by M.M. Golomazov (1966) [22].

## 3. The Method of Characteristics

### 3.1. Development of the Method of Characteristics

The method of characteristics as applied to the numerical integration of partial differential equations of hyperbolic type was first proposed by Massau (1900) over 70 years ago. This method has been widely used for the solution of a variety of problems in physics and mechanics.

The method of characteristics was redeveloped by specialists in gas dynamics, primarily for the calculation of steady two-dimensional and unsteady one-dimensional gas flows. The graphical method of characteristics proposed in the twenties by Busemann for the calculation of steady plane supersonic gas flows with constant entropy is well known. An analytic method of characteristics for axisymmetric flows with shock waves was developed by P.I. Frankl (Kochin, Kibel and Roze, 1963). Subsequently, the method of characteristics was generalized to other cases of gas flows and various modifications were proposed. Application of the classical method of characteristics for solving problems in gas dynamics without the use of electronic digital computers is described in many books and papers (see, for example, the well known monographs by R. Courant and K. O. Friedrichs

(1948) and A. Ferri (1949, 1954), A.A. Dorodnitsyn (1957)). The computational scheme conceived in this work laid the foundation for the numerical method of characteristics in its present form.

The advent of electronic computing machines made possible a sharp increase in the number of node points in the characteristic net and consequently permitted a significant increase in accuracy. At the same time, computers made it possible to apply the numerical method of characteristics in its most general form for the calculation of supersonic gas flows with physical-chemical processes. Let us consider now the development of the computer-oriented method of characteristics for calculating steady two-dimensional flows, a case that has been developed in detail.

The application of electronic computers requires that the computational schemes for the method of characteristics be expressed in a specific form. Nonetheless, this method was initially programmed for machine execution without modification. In particular, all of the known variants of the method of characteristics made use of a large number of trigonometric functions, the calculation of which occupied a significant portion of machine time. In 1959, the method of characteristics was modified by Ehlers (1959) who introduced new unknown functions, making possible a significant reduction in the number of elementary functions used. Ehler's method, however, was not applicable for a series of reasons (for example, because of the divergence of the iteration process when calculating points on a shock wave) in the general case. Further development of the method of characteristics was accomplished by P.I. Chushkin (1960) who utilized Ehler's valuable concept and proposed a working form of the method. A systematic exposition of the definitive variant of the numerical method of characteristics for steady two-dimensional supersonic flows of a perfect gas with constant specific heat was presented in a paper by O.N. Katskova, I.N. Naumova, Y.D. Shmyglevskii and N.P. Shulishnina (1961) which presents the necessary formulas and schemes of computation for a series of typical cases (flow in Laval nozzles, the flow about bodies of various shapes).

Recently the numerical method of characteristics was generalized to the case of supersonic flow of a gas with arbitrary thermodynamic properties. For equilibrium and non-equilibrium flows a working form of the method of characteristics convenient for execution on electronic digital computers was developed by O.N. Katskova and A.N. Kraiko (Katskova 1961; Katskova and Kraiko, 1962, 1963, 1964). Various schemes of constructing a computational mesh for the three-dimensional method of characteristics were proposed in papers by C. Ferrari (1949), R. Sauer (1950), A. Ferri (1954), V.V. Rusanov (1963), M. Holt (1956), D.S.Butler (1960) and P.I. Chushkin. (1968) [18 - 20] ; [23 - 25] .

As is well known, the method of characteristics

has certain advantages, as compared to other numerical methods- the pertinent equations undergo a significant simplification on the characteristic surfaces; furthermore, the method of characteristics admits considerable mathematical rigor (uniqueness and convergence have been proved). These circumstances have assured the wide-spread application of the numerical method of characteristics to the solution of two-and three-dimensional hyperbolic problems.

Some words about the properties of the characteristics. As is known, the hyperbolic weak disturbances spread in the flow along definite lines (surfaces) called characteristics. These are the Mach lines in the two-dimensional flow and the surfaces of conic type (conoids) in the three-dimensional flow. The condition that some line (surface) is a characteristic is the condition of impossibility of single-valued determination of the derivatives of all the unknown functions on this line (surface). This is due to the impossibility of the solution of the Cauchy problem with initial data on this surface. It may be shown, that some linear combination of the input equations may result in the fact that these input equations will contain only the interior derivatives along characteristic surfaces and will not contain the derivatives leading beyond the characteristic variety.

The use of this mathematically equivalent system for the construction of a numerical algorithm has certain advantages as compared to other numerical methods, namely, the solution of the equations becomes much simpler on the characteristic surfaces; using the characteristic network the range of dependence of the solution may be taken into account with great precision; in addition, the method of characteristics is mathematically rigorous (existence of the solution and convergence are proved for this method). These circumstsnces provide a rather wide application of the numerical method of characteristics for the solution of hyperbolic problems.

In constructing the numerical schemes of the method of characteristics in two- and three-dimensional problems both purely characteristic schemes (the range of integration is covered by the curvilinear characteristic network) and the numerical schemes allowing to calculate "with respect to layers" are used. We shall try to describe these approaches below.

Start with the account of the two-dimensional method of characteristics for the solution of steady supersonic problems in gas dymamics in the general case of non-equilibrium flows. [18]

## 1. Non-Equilibrium flows

Let us examine a steady plane ( $j = 0$ ) and axisymmetric ( $j = 1$ ) supersonic non-equilibrium flow of an inviscid and non-heat-conducting gas. Flow of this kind is described by the usual equation of continuity, motion and energy, relating velocity $V$, density $\rho$, pressure $p$, and specific enthalpy $h$.

Let the thermodynamic state of the gas be determined by the pressure $p$, the temperature $T$ of translational degrees of freedom of some component of the gas and $n$ parameters, $q(q_1, \ldots, q_n)$, characterizing irreversible processes. The $q$ parameters may be mass concentrations, energies of the internal degrees of freedom, and so on. Let us postulate that the variation in the parameters $q$ is represented by the following equations:

$$\frac{dq_i}{dt} = F_i(p,Tq) = \varphi^i(p,T,q)\, f_i(p,T,q) \qquad (i = 1,\ldots,n), \tag{3.1.1}$$

where $F_i$, $\varphi^i$ and $f_i$ are known functions of $p$, $T$, and $q$, the functions $\varphi^i$ being related to the speeds of physical-chemical processes, in particular when $\varphi^i \equiv 0$ the corresponding $q_i$ are frozen, while at $\varphi^i \equiv \infty$ they are in equilibrium. For equilibrium, the values of the parameter $q_i$ are determined by the equation

$$f_i(p,T,q) = 0 \tag{3.1.2}$$

Eqs. (3.1.1) encompass a wide range of processes, chemical reactions, excitation of internal degrees of freedom, variation in the translational temperatures of various components of the gas, and so on.

Let the equation of state of the gas also be known

$$\rho = \rho(p,T,q) \tag{3.1.3}$$

and the expression for specific enthalpy as well

$$h = h(p, T, q) \tag{3.1.4}$$

The solution will be carried out in Cartesian coordinates $x, y$. Relevant quantities will be expressed in dimensionless form, referring them to some linear dimension, velocity $V_\infty$ and density $\rho_\infty$ (in external flow problems, these may be the velocity and density of the incident flow) and also the gas constant of some gas $R_\infty$. In reducing the $q$ parameters to dimensionless form, their physical sense should be born in mind.

Let us postulate as the fundamental unknown functions the quantities

$$\zeta = \tan\Theta\,, \quad \beta = \sqrt{\frac{V^2}{a^2} - 1}\,, \tag{3.1.5}$$

where $\Theta$ is the angle of inclination of the velocity vector to the $x$ axis*, and $a$ is the frozen velocity of sound determined by the expression

$$a^{-2} = \rho_p + \frac{\rho_T}{h_T}\left(\frac{1}{\rho} - h_p\right). \tag{3.1.6}$$

Here and below the latter subscripts of $\rho$ and $h$ refer to the appropriate partial derivatives.

In the case of supersonic flow there are three families of real characteristics: two families of Mach lines (characteristics of the first and second families) and the stream-

*It is assumed that in the flow region being computed $\Theta$ is not equal to $\pm\pi/2$. If the contrary is true, then instead of the function $\zeta$ the function $\bar{\zeta} = \cot\Theta$ is introduced.

lines ($\Psi$ = constant).

The characteristics of the first (using the lower sign in the formulas) and of the second (the upper sign) families are determined by the relationships

$$\text{(3.1.7)} \qquad dy = \frac{\beta\zeta \pm 1}{\beta \mp \zeta}\, dx = B^{\pm} dx$$

$$\text{(3.1.8)} \qquad d\Psi = \pm \frac{\rho V y^{j} (1+\zeta^2)^{1/2}}{\beta\zeta \pm 1}\, dy$$

$$\text{(3.1.9)} \qquad \frac{1}{1+\zeta^2}\, d\zeta \pm \frac{\beta}{\rho V^2}\, dp \pm \frac{1}{\beta\zeta \pm 1}\left[ j\frac{\zeta}{y} - \frac{(1+\zeta)^{1/2}(\rho_T h_{qi} - h_T \rho_{qi}) F_i}{h_T \rho V} \right] dy = 0,$$

tensor notation being introduced into the last equation with summation over $i$ understood for repeated lower indices or subscripts. Let us note that in applying eqs. (3.1.7) to (3.1.9) special cases may arise when the angle of inclination of the characteristics of the first and second families either tend toward $\pi/2$ (that is, $\beta-\zeta \to 0$ and $\beta+\zeta \to 0$) or else tend to zero (that is, $\beta\zeta+1 \to 0$ and $\beta\zeta-1 \to 0$). In these cases multiplying eqs. (3.1.7) to (3.1.9) by appropriate factors all possible singularities may be eliminated.

Along the streamlines the following relations are satisfied:

$$\text{(3.1.10)} \qquad dq_i = \frac{F_i (1+\zeta^2)^{1/2}}{V}\, dx = C_i\, dx$$

$$dT + D\,dp + E_i\,dq_i = 0, \qquad (3.1.11)$$

where

$$D = \frac{\rho h_p - 1}{\rho h_T}, \quad E_i = \frac{h_{qi}}{h_T}, \qquad (3.1.12)$$
$$\frac{V^2}{2} + h = h_0(\Psi),$$

where $h_0(\Psi)$ is the total enthalpy of the gas, a known function of $\Psi$. In using the method of characteristics it is necessary to solve a series of elementary problems. In this method the point in question lies at the intersection of lines determined either by differential or finite equations. These lines may be, for example, characteristics, streamlines or shock waves. The unknown functions along these lines are also determined by either differential or finite relationships.

We shall make use of the method of characteristics, in which the differential equations of the form $\sum_i A_i dz_i = 0$, holding on a certain line ab, are replaced by finite difference equations of second order accuracy

$$\frac{1}{2}\sum_i (A_{ia} + A_{ib})(z_{ib} - z_{ia}) = 0.$$

Since the resulting system of finite equations cannot be solved explicitly, it is solved numerically by transforming it to a form convenient for solution by successive approximations.

Now we shall examine two approaches to the method of characteristics. In the first the characteristic net is constructed in the course of the calculations as a series of intersections of characteristics of two families, that is, the $x$, $y$ coordinates of the node points are calculated simultaneously along with the parameters of the flow. In the second approach the calculation is carried out in layers, bounded by lines $x_0$ and $x_0+\Delta x$, while within each layer the segments of the characteristics are so placed that at $x_0+\Delta x$ they pass through points characterized by some given value of $y$.

We shall now describe the solution of several elementary problems, utilizing the first of the two approaches to the method of characteristics [18, 19].

a) Let us consider a case in which the unknown point lies within the field of flow. Let the flow parameters be known on some non-characteristic line, in particular at the

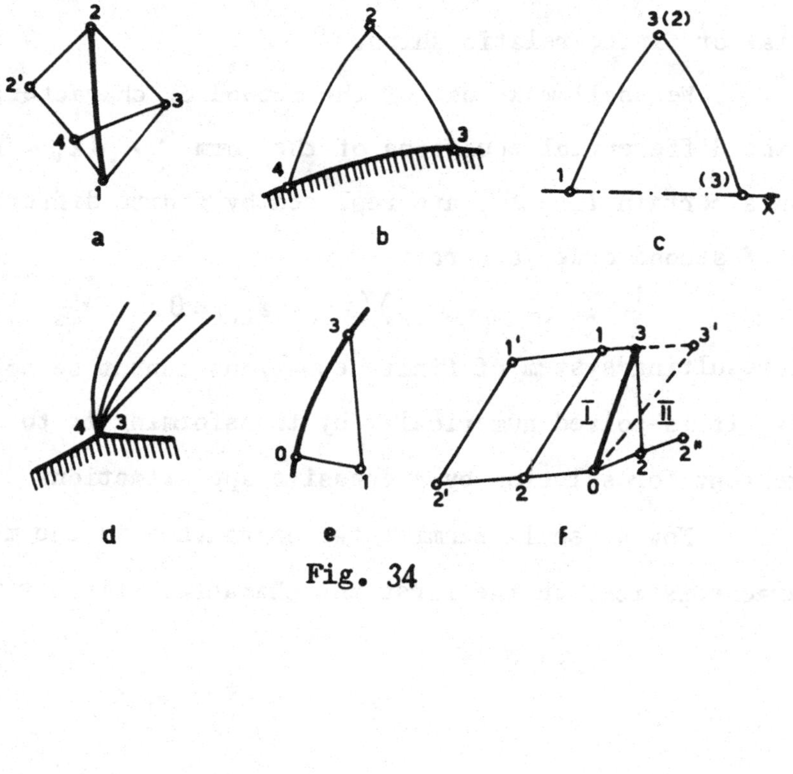


Fig. 34

points 1 and 2 (Fig. 34 a - f), and let us determine the flow parameters at the point 3 lying at the intersection of the characteristic of the first family 1-3, with the characteristic of the second family 2-3. Functional values at the points 1,2,... will be designated by the appropriate subscript.

Let us write eqs. (3.1.7) to (3.1.9) along the characteristics 1-3 and 2-3 in finite difference form and solve them for $y_3, x_3, \Psi_3, p_3, \zeta_3$. In these equations the averaged coefficients of the differences consist of two components, the first of which is known. Substituting the first component for the second, the above five quantities are determined in the first approximation. Knowing $\Psi_3$ one can determine the point of intersection 4 of the streamline passing through point 3, and the known characteristic segment 1-2. The parameters at point 4 are determined by quadratic interpolation for the value $\Psi_4 = \Psi_3$ from the known data on the characteristic 1-2. Then, after having written eqs. (3.1.10) and (3.1.11) along the segment of streamline in finite difference form we obtain

$$q_{i3} = q_{i4} + C_i (x_3 - x_4) \qquad (3.1.13)$$

$$T_3 = T_4 - D(p_3 - p_4) - E_i(q_{i3} - q_{i4}). \qquad (3.1.14)$$

The coefficients in (3.1.13) to (3.1.14) also consist of two components. Substituting the first component for the second we find $q_{i3}$ and $T_3$ in the first approximation. Finally, on the basis of the quantities already found, we determine $\rho_3, h_3, V_3, \alpha_3$ and $\beta_3$ from eqs. (3.1.3) to (3.1.6).

This process must then be repeated. In each new approximation the second terms in the averaged coefficients are determined as a result of the preceding approximation. Three or four such iterations are generally sufficient for satisfactory accuracy.

Unfortunately, in approaching equilibrium, the convergence of this method deteriorates, requiring a significant decrease in the integration step, which increases the time of the calculation enormously. As shown analytically (Katskova and Kraiko, 1963, 1964) this is related to the finite difference relationships (3.1.13) which contain the values of $f_i$ at the unknown point 3. Close to equilibrium the quantities $f_i$ tend to zero and are extremely sensitive to changes in $T$ and the corresponding parameters $q_i$. Consequently, a small error in $T_3$ and $q_{i3}$ in the previous approximation results in a large error in $q_{i3}$ in the succeeding approximation. This in turn leads to an increased error in the computed temperature $T_3$ which is found using eq. (3.1.14) from $q_{i3}$, and so on.

b) Next we shall consider the case when the unknown point lies on a solid boundary (Fig. 34b). Taking account of the fact that point 3 lies at the intersection of a characteristic of the second family 2-3 and the boundary 4-3, the equation of which has the form $y = y_b(x)$, we obtain

$$x_3 = \frac{y_b(x_3) - x_3 y_b'(x_3) - y_2 + B^- x_2}{B^- - y_b'(x_3)}, \tag{3.1.15}$$

$$y_3 = y_b(x_3), \quad \zeta_3 = y_b'(x_3),$$

for which, in the first approximation, the quantity $x_3$ in the right hand side of the equation for $x_3$ is replaced by the coordinate of the preceding point $x_4$. By writing eq. (3.1.9) in finite difference form along the characteristic of the second family we obtain an expression for $p_3$ in terms of $\zeta_3$ and $y_3$. In view of the fact that the solid boundary is a streamline, $\Psi_3 = \Psi_4$. The remaining parameters are determined as before by successive approximation.

c) In calculating axisymmetric flows ($j = 1$) special attention must be given to points on and near the axis. This is a result of the indeterminate nature of the ratio $\zeta / y$ entering eq. (3.1.9).

Suppose the known point 1 lies on the axis and suppose we wish to determine the parameters at a neighboring point 3 (Fig. 34 c). To eliminate the indeterminate form the relationship (3.19) is multiplied by $y$. The resulting equation then replaces eq. (3.1.9) and is written in finite difference from along the characteristic of the first family 1-3, which leads to formulas for the unknown parameters similar to the formulas for a point in the interior of the flow field.

Wherever point 3 lies on the axis of symmetry $y_3 = \zeta_3 = 0$. If this is taken into account and eqs. (3.1.7) and the transformed equation (3.1.9) are written in finite difference form along the characteristic of the second family 2–3 we obtain expressions for $x_3$ and $p_3$.

d) Certain new features arise in connection with calculating corner points on a solid boundary (Fig. 34d) and focal points of the characteristics. Here it becomes necessary to determine the parameters at points having identical coordinates in the plane of flow but corresponding to distinct characteristics of one family. Let us denote parameters at the known point by the subscript 4 and at the unknown point by the subscript 3. It is evident that:

(3.1.16) $\quad x_3 = x_4 \;, \; y_3 = y_4 \;, \; \Psi_3 = \Psi_4 \;, \; q_{i3} = q_{i4} \;.$

Eqs. (3.1.9) and (3.1.11) written in finite difference form permit $p_3$ and $t_3$ to be expressed in terms of $\zeta_3$. The quantities $\rho_3$ and $h_3$ are subsequently found from eqs. (3.1.3) and (3.1.4). In this fashion all of the parameters at point 3 are determined by the selection of $\zeta_3$.

e) Finally, let the unknown point 3 lie at the intersection of a shock wave 0–3 with a characteristic of the first family 1–3 (Fig. 34e). The incident flow and the flow parameters at the point 0 and 1 are assumed known.

Because of the vanishing thickness of the shock

wave we have in accordance with eqs. (3.1.1) $q_{i3} = q_{i\infty}$ (here the index $\infty$ refers to parameter values in the incident flow). From these equalities and eqs. (3.1.1), (3.1.3), (3.1.4) the quantities $\rho_3$, $h_3$ and $F_{i3}$ are seen to be functions of $P_3$ and $T_3$ alone. With a uniform incident flow at velocity $V_\infty$ the shock wave relation to determine the dimensionless quantities in question may be written in the form (Naumova, 1963)

$$V_3^2 = \frac{1}{1+\tau_3^2}\left(\frac{\tau_3^2}{\rho_3^2}+1\right), \quad \zeta_3 = \frac{\tau_3(\rho_3-1)}{\rho_3+\tau_3^2},$$

$$\Delta_1 \equiv \frac{\tau_3^2}{1+\tau_3^2}\left(1-\frac{1}{\rho_3}\right)+p_\infty - p_3 = 0, \tag{3.1.17}$$

$$\Delta_2 \equiv \frac{1}{2} + h_\infty - \frac{V_3^2}{2} - h_3 = 0$$

where $\tau$ is the tangent of the angle of inclination of the shock wave to the $x$-axis.

Then writing the equation of the shock wave $dy = -\tau dx$ and eqs. (3.1.7) in finite difference form along a characteristic of the first family 1-3, we obtain formulas for determining the coordinates $x_3, y_3$ at the point of intersection of shock wave and characteristic. The finite difference analogue of eq. (3.1.9) for the characteristic 1-3 can be represented in the form

$$\Delta_3 = 0, \tag{3.1.18}$$

where $\Delta_3$ is expressed in terms of functional values at the points 1 and 3.

The equations and relationships for the shock wave listed here form the complete system of equations for determining all the parameters at point 3. To solve these equations in the first approximation the quantities $\tau_3, p_3, T_3$ close to $\tau_0, p_0$ and $T_0$ are given. This makes it possible to determine the remaining parameters and also quantities $\Delta_1, \Delta_2, \Delta_3$, which, generally speaking, differ from zero. It is necessary to find such values of $\tau_3, p_3$ and $T_3$ for which the equalities $\Delta_1 = \Delta_2 = \Delta_3 = 0$ will hold. This is achieved through proper selection and interpolation. The quantity $\Psi_3$ equals the value $\Psi$ in the accumulation flow, that is $\Psi_3 = y_3^{j+1}/(j+1)$ .

## 2 Perfect Gas

For a perfect gas with constant adiabatic exponent, these relationships are such that analytic expressions for $p, T,\ \ , h$ and $V$ may be obtained in terms of $\beta$ and $S$. Let us now make use of these expressions in eliminating $\varrho$, $V$ and $p$ from eqs. (3.1.8) and (3.1.9) (the reader is reminded that now $h_{q_i} dq_i = 0$) and introduce for $S \neq \text{const.}$ a new generalized stream function $\bar{\Psi}$, referred to as $a_{cr}$ (rather than $V_\infty$) and related to the usual stream function by the relationship [19], [23]

(3.1.19)
$$d\bar{\Psi} = \gamma^{\frac{1}{\gamma-1}}\, e^{\frac{S}{\gamma-1}}\, d\Psi$$

Then eqs. (3.1.8) and (3.1.9) assume the form

$$d\zeta \mp \frac{2\beta^2(1+\zeta^2)}{(\gamma+1)(1+\beta^2)(1+K\beta^2)}\, d\beta \pm j\, \frac{\zeta(1+\zeta^2)}{(\beta\zeta \pm 1)y}\, dy \mp \frac{\beta(1+\zeta)}{\gamma(\gamma-1)(1+\beta^2)}\, dS = 0$$

(3.1.20a)

$$d\bar{\Psi} \mp \frac{y^{i}\sqrt{(1+\beta^2)(1+\zeta^2)}}{(\beta\zeta \pm 1)(1+K\beta^2)^{1/2K}}\, dy = 0 \tag{3.1.20b}$$

$$\left(K = \frac{\gamma-1}{\gamma+1}\right)$$

Writing eqs. (3.1.7), and (3.1.20) in terms of finite differences and bearing in mind the $S = S(\bar{\Psi})$ we obtain a system of equations for determining the five unknowns $x_3$, $y_3$, $\Psi_3$, $\zeta_3$, $\beta_3$. Thus for a perfect gas, in contrast with the general case, the number of unknowns has been reduced by three. Significant simplifications also take place in the formulas for computing points on the shock wave; in solving the system one needs no longer select three quantities but only one, the tangent of the angle of inclination of the shock wave $\tau$ .

It should be emphasized, however, that these simplifications leading to substantial economies in machine time, yield equations lacking generality, to a certain extent, because of an implied specific form of the thermodynamic functions. Computer programs may be compiled more conveniently when different gases are considered merely by substituting an appropriate program for computing the required thermodynamic functions.

In concluding this section, we shall pause to consider the application of the numerical method of characteristics along strips bounded by lines $x$ = constant. This scheme for machine calculations of axisymmetric supersonic flows of a perfect gas with shock waves was developed by O.N. Katskova and P.I. Chushkin (1964). A scheme of characteristics along strips (but only for the isentropic case) in unsteady one-dimensional problems, was proposed earlier by M. Lister (1960). S.K. Godunov and K.A. Semedayev (Godunov, 1962) also investigated a scheme of this type in application to unsteady one-dimensional gas flows. They did not require, however, that along the new strip $x_0 + \Delta x$ the characteristics would pass through points with a given value [20].

While the method of characteristics along strips may, by analogy with the usual scheme, be extended to the case of a real gas in either equilibrium or non-equilibrium, we shall, for simplicity, limit the present discussion to the case of a perfect gas.

We shall discuss the method of characteristics along strips in relation to a concrete example, the calculation of the supersonic flow about a body of revolution the contour of which is given by the equation $y = y_b(x)$. The solution will be obtained in the variables $x, \xi = (y - y_b)(y_s - y_b)$ where $y = y_s(x)$ is the unknown equation of the shock wave.*

---

*For calculating the flow in a nozzle the pertinent variables will be $x, \xi = y \;\; y_b$ where now $y = y_b(x)$ is the equation of the nozzle wall.

Utilizing as before the fundamental unknown functions $\zeta, \beta$ and $S$, the variable $y$ is eliminated from the differential equations of the characteristics (3.1.7) and (3.1.20a) by the formulas

$$y = \varepsilon\xi + y_b \; , \quad dy = \varepsilon d\xi + \lambda dx$$

where

$$\varepsilon = y_s - y_b \; , \quad \lambda = (y_s' - y_b')\xi + y_b'$$

The following equation replaces eq. (3.1.20b) for the streamline

$$\frac{d\xi}{dx} = \frac{1}{\varepsilon}(\zeta - \lambda) \, . \qquad (3.1.21)$$

We shall now indicate the order of the calculations. Let there be known on some strip boundary $x = x_0$ lying between the shock wave ($\xi = 1$) and the body ($\xi = 0$) in a supersonic region of flow, the values of the fundamental functions at a series of points $n$ corresponding to $\xi = \xi_n$ = constant. To find the values of these functions on the next strip boundary $x_0 + \Delta x$, one must solve elementary problems by schemes analogous to the usual ones (as applied to the case of a perfect gas but involving equations written in $x, \xi$ coordinates).

The calculations are begun at a known point 3 on the shock wave (Fig. 35a) where the tangent of the angle of the shock $\tau$ is selected to satisfy the compatibility relationship of the characteristic of the first family 1-3. In computing at a point 3 within the field of flow (Fig. 35b) characteristics of the first family 1-3 and of the second family 2-3 and the streamlines 3-4 are extended outward from the point. In order to calculate the single unknown $\beta_3$ at point 3 on the body contour one makes use of appropriate finite difference equations along the characteristic of the second family 2-3 (Fig. 35c). In all of these cases the values of the fundamental functions in the points 1,2 and 4 on the layer $x_0$ are found by quadratic interpolation between the points $n-1$ , $n$ and $n+1$. The magnitude of the step $\Delta x$ must be coordinated with the step $\Delta\xi$ and is selected on the basis of accuracy and stability of the computations.

Let us note that the method of characteristics along strips, while requiring rather more machine time, posseses a number of advantages compared to the usual scheme. In this approach the coordiantes of the node points are known at the start and consequently need not be retained in the machine memory. The computational network in this case can be easily changed in the process of calculation, which permits it to be chosen in the most favourable manner. In this system the calculated parameters are obtained on successive sections $x =$ constant at a series of present points, which is also needed in practice. One can, furthermore, make use of a certain extrapolation process for increasing the accuracy of the calculations (Lister, 1960). Finally, the method of characteristics along strips can, as has already been mentioned, be combined with the finite difference method for solving three-dimensional problems. Some cal-

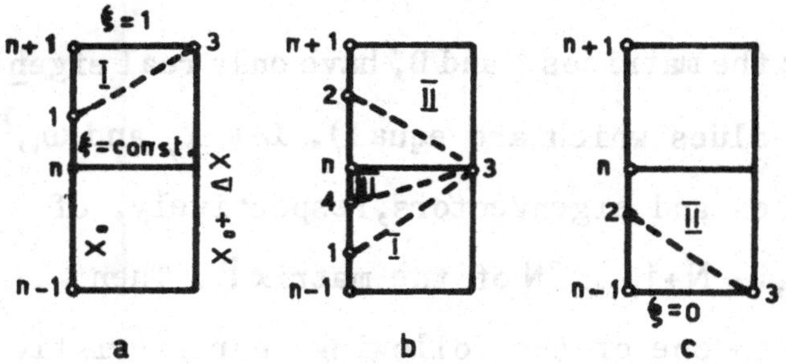


culated results you can find in [24] .

Fig. 35

## 3.2. Construction of Difference Schemes for Equations of hyperbolic Type on the Basis of Characteristic Relations

In this section we give a scheme for constructing a characteristic network method developed by K. Magomedov and A. Kholodov [25] .

### 1. Construction of the Difference Scheme for the System of Quasilinear Equations of Hyperbolic Type with Respect to Many variables

Consider the system of quasilinear equations of hyperbolic type of the first order in the general form

(3.2.1) $$u_t + Au_x + Bu_y + \dots = f \, ,$$

where the matrices A and B, the column vector on the right-hand sides, f, and the vector function, $v$, may be functions of $t, x, y, \dots$, the unknown variables with N components. We shall further restrict to the terms of Eq. (3.2.1) i.e. consider the three-dimensional case as the extension of the method to a greater number of variables is formal.

**Suppose that the matrices A and B, have only real eigen values (there may be some values which are equal). Let $\mu_i$ and $\omega_i$,** $i = 1,2,\ldots, N$, be eigenvalues and eigenvectors, respectively, of the matrice $A'$ and $\mu_i$ and $\omega_i$, $i = N+1,\ldots,2N$ of the matrix $B$. Then Eq. (3.2.1) may be reduced to one of the following characteristic (normal) forms

(3.2.2) $$\omega_i u_{ti} = \omega_i f - \omega_i (Bu_y), \quad i = 1,\ldots, N,$$

(3.2.3) $$\omega_i u_{ti} = \omega_i f - \omega_i (Au_x), \quad i = N+1,\ldots, 2N .$$

Here $u_{ti} = \partial u/\partial t + \mu_i \partial u/\partial x$, $i=1,\ldots,N$; $u_{ti} = \partial u/\partial t + \mu_i \partial u/\partial y$, $i=N+1,\ldots,2N$, are derivatives along two-dimensional characteristics on the surfaces $y = \text{const}$ and $x = \text{const}$, respectively, or, in other words, along the lines of intersection of characteristic and coordinate surfaces. Really Eq. (3.2.2), for example, may be derived in the following way. Multiply scalarly Eq. (3.2.1) by $\omega_i (\omega_i^2 \neq 0)$ and use the equalities $u_t = \partial u/\partial t = u_{ti} - \mu_{ti} \partial u/\partial x$.

Then

(3.2.4) $$\omega_i (u_{ti} - \mu_i u_x) + \omega_i (Au_x) = \omega_i f - \omega_i (Bu_y)$$

As $\omega_i (Au_x) = (A'\omega_i) u_x$, where $A'$ is the conjugated matrix $A$, then $\omega_i u_{ti} + (A'\omega_i - \mu_i \omega_i) u_x = \omega_i f - \omega_i (Bu_y)$. By the assumption, $\mu_i$ and $\omega_i$ are the eigenvalues and eigenvectors of the matrix $A'$, therefore $A\omega_i - \mu_i \omega_i = (A' - \mu_i E)\omega_i = 0$ **and equations (3.2.2) are valid. Eqs. (3.2.3) are derived analogously. Each equation of the system has derivatives only in two directions.**

**Consider now the 3N Eqs. (3.2.1), (3.2.2), (3.2.3)**

simultaneously. It is obvious that the number of independent variables among them is equal to $N$. We can use this fact to construct explicit difference schemes not requiring, for example, any approximation of partial derivatives with respect to $x$ and $y$. If we know the solution at the point $H$ on the layer $t = t_0 = \text{const}$ we can find the solution at the point $H$ on the layer $t = t_0 + \tau$ by writing down eqs. (3.2.1), (3.2.2), (3.2.3) in difference form at $t = t_0 + (1-\nu)\tau$, $0 \le \nu \le 1$, i.e. the coefficients and partial derivatives with respect to $x$ and $y$ will be found with this value of $t$; then we shall obtain

$$\begin{aligned} &u(H) - u_0 + \tau\nu (Au_x + Bu_y - f)_0 + \tau(1-\nu)\left[Au_x(H) + Bu_y(H) - \right.\\ &\qquad \left. - f(H)\right] = 0\left[(1-2\nu)\tau^2 + \tau^3\right], \\ &\omega_i\left[u(H) - u_i\right] + \tau\nu\,\omega_i (Bu_y - f)_i + \\ &\quad + \tau(1-\nu)\omega_i\left[Bu_y(H) - f(H)\right] = 0\left[(1-2\nu)\tau^2 + \tau^3\right],\ i = 1,\dots,N, \\ &\omega_i\left[u(H) - u_i\right] + \tau\nu\omega_i (Au_x - f)_i + \tau(1-\nu)\omega_i\left[Au_x(H) - f(H)\right] = \\ &\qquad = 0\left[(1-2\nu)\tau^2 + \tau^3\right],\ i = N+1,\dots,2N. \end{aligned} \tag{3.2.5}$$

Here the indices $i$ and 0 designate the values of the functions at the points of intersection of the characteristics and the line $x = \text{const}$, $y = \text{const}$ drawn from the point $H$ with the surface $t = t_0$. The tangential vectors of these characteristics have the form $\{1, \mu_i, 0\}$ at $i = 1,\dots,N$ and $\{1, 0, \mu_i\}$ at $i = N+1,\dots,2N$.

Eqs. (3.2.5) contain $3N$ unknowns $u(H), u_x(H), u_y(H)$ and the same number of equations. We can show that by elimination of $u_x(H), u_y(H)$ from this system we find $u(H)$.

Actually, designating $\tau(1-\nu)Au_x(H) = V, \tau(1-\nu)Bu_y(H) = W$ we shall have from Eqs. (3.2.5) $u(H) + V + W = F_1[u(H), G]$,

(3.2.6) $$\Omega_1[u(H) + W] = F_2[u(H), G], \Omega_2[u(H) + V] = F_3[u(H), G],$$

here $G = G[t_0, x, y, u(t_0, x, y), \nu\tau u_x(t_0, x, y), \nu\tau u_y(t_0, x, y)]$, $\Omega_1$ and $\Omega_2$ are the matrices whose rows are, respectively, the coordinate vectors $\omega_i$, $i = 1, \dots, N$ and $\omega_i$, $i = N+1, \dots, 2N$. As the eigenvectors have been chosen linearly independent, then the matrices $\Omega_1$ and $\Omega_2$ have inverse matrices $\Omega_1^{-1}$ and $\Omega_2^{-1}$. By elimination $V$ and $W$ from Eqs. (3.2.6) we have the difference equations approximating Eq. (3.2.1):

(3.2.7) $$u(H) = \Omega_1^{-1}F_2 + \Omega_2^{-1}F_3 - F_1 + O[(1-2\nu)\tau^2 + \tau^3].$$

In deriving this relation finite-difference relations simply replace ordinary derivatives along certain lines. The conservation in the difference equations of partial derivatives with respect to $x$ and $y$ on the layer $t = t_0$ depend, as it is seen from Eq. (3.2.5), on the choice $\nu$. If $\nu = 1/2$ the Eqs. (3.2.7) will be correct to the second order of approximation with respect to $t$, and, generally speaking, contain $u_x(t_0, x, y), u_y(t_0, x, y)$. If $\nu = 0$ (the first order approximation with respect to $t$) these derivatives are dropped. Apparently, in constructing difference schemes one should avoid, if possible, numerical differentiation of the known functions, for example, on the layer $t = t_0$. For a special kind of equations (3.2.1), i.e. gas dynamics equations, a more complicated way of deriving difference equations of type (3.2.7) is related to the use of certain bicharacteristics (unlike the lines considered here they are not situated on the coordinate surfaces). It was suggested in [25] where the partial derivatives on the layer $t=t_0$ are retained and in [24] where these derivatives are replaced by derivatives with respect to the bicharacteristics.

Thus equations (3.2.7) express the unknown functions at the point H by means of their values on the layer $t=t_0$ By introducing the fixed network $t=n\tau$ $(n=0,1,2,\dots)$, $x=mh_1$, $y=lh_2$, where $m,l$ take integral values according to the region considered, and denoting the values of functions at the nodes of the network by $u^n_{ml}$ and, furthermore, given the relation of the parameters with indices 0 and $i=1,2,\dots,2N$ with known values of $u^n_{ml}$ on each layer $t=t_0=n\tau$, we arrive at the ordinary difference scheme with fixed network or the method of networks

$$u^{n+1}_{ml} = \sum_{ij} c_{ij} u^n_{m+i,l+j} + \tau F(t_0, t_0+\tau), \qquad (3.2.8)$$

where $c_{ij}$ are matrices whose elements for the schemes of the first order depend only on the parameters for $t=t_n$ and in the second order schemes for quasilinear equations also on $u^{n+1}_{ml}$. Most frequently, in constructing numerical methods using the characteristic relations importance is attached to the derivation of equations of type (3.2.7), not to the choice of nodes and interpolation, whereas it is the latter that substantially affect approximation and stability of the difference scheme. The difference schemes derived from Eqs. (3.2.7) for linear (in this case we may assume $\nu = 0$ ) and quadratic interpolation are of the greatest interest. Only the latter may be used since in this case we shall have the scheme of the second order with respect to all variables. At quadratic interpolation, as will further be seen, most frequently results in schemes stable only at $\tau \sim O(h^2)$ , whereas the necessary condition of stability of the CFL condition for hyperbolic equations has the form $\tau \sim O(h)$.

Proceed to the study of the above scheme for elementary equations with constant coefficients.

## 2 Investigation of the Difference Scheme for Model Equations

Consider Eq. (3.2.1) at $N = 1$ . Then $A$ and $B$ are scalar quantities. The characteristic equation in the difference form (3.2.7) will have the form

$$u(H) = u_1 + u_2 - u_0 - \tau\nu\left[(Au_x)_2 + (Bu_y)_1 - (Au_x)_0 - (Bu_y)_0\right] + \tag{3.2.9}$$
$$+ \tau\nu(f_1 + f_2 - f_0) + \tau(1-\nu)\, f(H) + O\left[(1-2\nu)\tau^2 + \tau^3\right].$$

Without losing generality one may consider that $A > 0$ and $B > 0$ . $u_1$ and $u_2$ designate the values of functions at the point $H$ with the plane $t = \text{const}$.

By linear interpolation we have the scheme

$$u^{n+1}_{ml} = u^n_{ml}(1-\alpha_1-\alpha_2) + \alpha_1 u^n_{m-1,l} + \alpha_2 u^n_{m,l-1} + \tau f^n_{ml} + O\left[\tau^2(1+\alpha_1^{-2}+\alpha_2^{-2})\right].$$

(3.2.10)

Here $\alpha_1 = A\tau/h_1, \alpha_2 = B\tau/h_2$. By fulfilling the CFL condition it is

easy to check the stability of this cheme using the maximum principle, since the index of the difference scheme
$I = |1-\alpha_1-\alpha_2| + \alpha_1 + \alpha_2 + c\tau < 1 + c\tau$ at $\alpha_1+\alpha_2 < 1$ where $c = \max\limits_{m,l} |f^n_{ml}|$.

With quadratic interpolation and the scheme is stable only at $\alpha_i^2 = 0(\tau)$. For $\nu$ not equal to zero it is necessary to use an additional point. Take the point $(m-1, l-1)$. Then we have the scheme

$$\begin{aligned} u^{n+1}_{ml} &= (1-\alpha_1^2-\alpha_2^2+2\nu\alpha_1\alpha_2)u^n_{ml} + 2\nu\alpha_1\alpha_2 u^n_{m-1,l-1} - \alpha_1(1-\alpha_1)u^n_{m+1,l}/2 + \\ &+ \alpha_1(1+\alpha_1-4\nu\alpha_2)u^n_{m-1,l}/2 - \alpha_2(1-\alpha_2)u^n_{m,l+1}/2 + \\ &+ \alpha_2(1+\alpha_2-4\nu\alpha_1)u^n_{m,l-1}/2 + \\ &+ \nu\tau(f^n_1+f^n_2-f^n_0) + \tau(1-\nu)f^{n+1}_{ml} + 0\left[(1-2\nu)\tau^2+\tau^3+h_1^3+h_2^3\right]. \end{aligned} \tag{3.2.11}$$

The application of the method of separation of variables for the investigation of stability of the solution of the form $u^n_{ml} = \lambda^n \exp(imk_1h_1 + ilk_2h_2)$ of the homogeneous equation(3.2.11 shows that $|\lambda| < 1$ i.e. by fulfilling the CFL condition the scheme is stable only for $\nu = 1/2$. For the other values of $\nu$ the scheme is stable at $\alpha^2 = 0(\tau)$, $\max|\lambda| = 1 + \alpha_1^2 + \alpha_2^2$.

Using the Fourier method one may see that the use of additional nodes on the plane $t = n\tau$ deteriorates stability. For example, by fulfilling the CFL condition and using one more point linear interpolation $(m-2, l)$ renders the scheme unstable in general. By interpolation of high order and using new points we obtain schemes stable only at $\alpha_i^2 = 0(\tau)$.

We come to the same conclusions by analyzing acoustic equations in two and three dimensions which present a rather successful and simple model of gas dynamics equations. On the basis of these equations we shall associate the approach with the methods [25] . We shall examine the system

$$p_t + u_x + v_y = 0 , \quad u_t + p_x = 0 , \quad v_t + p_y = 0 . \tag{3.2.12}$$

In the two-dimensional case the difference equation (3.2.7) has the form

$$p(H) = \frac{p_1 + p_2}{2} - \frac{u_1 - u_2}{2} , \quad u(H) = \frac{u_1 + u_2}{2} - \frac{p_1 - p_2}{2} . \tag{3.2.13}$$

It is obvious that these relations give the accurate solution at the point $H$ on the layer $n+1$ by means of the values of $p$ and $u$ on the layer $n$ at any value of $\tau$ . Taking three nodes on the layer $n$ we have the difference scheme

$$\begin{aligned} p_m^{n+1} &= p_m^n + \frac{\alpha^j}{2}\left(p_{m+1}^n + p_{m-1}^n - 2p_m^n\right) - \frac{\alpha}{2}\left(u_{m+1}^n - u_{m-1}^n\right), \\ u_m^{n+1} &= u_m^n + \frac{\alpha^j}{2}\left(u_{m+1}^n + u_{m-1}^n - 2u_m^n\right) - \frac{\alpha}{2}\left(p_{m+1}^n - p_{m-1}^n\right), \end{aligned} \tag{3.2.14}$$

where $j = 1$ for linear interpolation and $j = 2$ for quadratic interpolation. These schemes are stable at $\alpha \le 1$, since the maximum eigenvalue of the conversion matrix in modulus $G(\tau, k)$ has the form

$$\max|\lambda|^2 = 1 - 4\alpha^3(1-\alpha^j)\sin^2\frac{kh}{2}$$

At $j = 1$ the scheme has approximation of the first order and it coincides with the numerical method, at $j = 2$ we obtain the scheme of second order approximation .

Consider now the three-dimensional case. Then the difference equation (3.2.7) will have the form

$$p(H)=\frac{p_1+p_2+p_3+p_4-2p_0}{2}-\frac{u_1-u_2}{2}-\frac{v_3-v_1}{2}+0\left[\nu\tau\left(h_1^2+h_2^2\right)+\tau^3\right],$$

$$u(H)=\frac{u_1-u_2}{2}-\frac{p_1-p_2}{2}+\tau\nu h_1 v_{xy}+0\left[\nu\tau h_1^2+\tau^3\right], \qquad (3.2.15)$$

$$v(H)=\frac{v_1-v_2}{2}-\frac{p_3-p_4}{2}+\tau\nu h_2 u_{xy}+0\left[\nu\tau h_2^2+\tau^3\right].$$

Taking five nodes one may obtain the difference schemes

$$p_{m\ell}^{n+1}=p_{m\ell}^{n}+\frac{\alpha_1^2}{2}\left(p_{m+1,\ell}^{n}+p_{m-1,\ell}^{n}-2p_{m\ell}^{n}\right)+\frac{\alpha_2^2}{2}\left(p_{m,\ell+1}^{n}+p_{m,\ell-1}^{n}-2p_{m\ell}^{n}\right)-$$
$$-\frac{\alpha_1}{2}\left(u_{m+1,\ell}^{n}-u_{m-1,\ell}^{n}\right)-\frac{\alpha_2}{2}\left(v_{m,\ell+1}^{n}-v_{m,\ell-1}^{n}\right),$$

$$u_{m\ell}^{n+1}=u_{m\ell}^{n}+\frac{\alpha_1^2}{2}\left(u_{m+1,\ell}^{n}+u_{m-1,\ell}^{n}-2u_{m\ell}^{n}\right)-$$
$$-\frac{\alpha_1}{2}\left(p_{m+1,\ell}^{n}-p_{m-1,\ell}^{n}\right)+\nu\alpha_1 h_1^2\left(v_{xy}\right)_{m\ell}^{n}, \qquad (3.2.16)$$

$$v_{m\ell}^{n+1}=v_{m\ell}^{n}+\frac{\alpha_2^2}{2}\left(v_{m,\ell+1}^{n}+v_{m,\ell-1}^{n}-2v_{m\ell}^{n}\right)-$$
$$-\frac{\alpha_2}{2}\left(p_{m,\ell+1}^{n}-p_{m,\ell-1}^{n}\right)+\nu\alpha_2 h_2^2\left(u_{xy}\right)_{m\ell}^{n}.$$

Here $j=1$ for linear interpolation between the adjacent points and $j=2$ for quadratic interpolation. When constructing the scheme of the second order additional nodes must be taken to determine the compound derivatives $u_{xy}$ and $v_{xy}$ in Eq. (3.2.16).

Resume the consideration of the Eqs.(3.2.16). The stability of different schemes may be examined by constructing the conversion matrix $G(\tau,k)$ for this scheme. Let $\nu=0$ . If $j=2$, then the corresponding scheme is unstable at $\alpha_1=\text{const}$, it is stable at $\alpha_1^2=0(\tau)$ and it is stable at $j=1$ by fulfilling the CFL condition. If $\nu=1/2$ and taking additional four points for the calculation of the compound derivatives the necessary Neumann condition is satisfied by fulfilling the CFL condition.

Let other schemes which are stable by fulfilling the CFL condition along one of the directions be possible, for example $\alpha_1<1$ , but it is required that $\alpha_2^2 = 0(\tau)$ along the other direction. In practical calculations by means of this scheme a step with respect to the variable was chosen depending on the character of the solution of the problem so that $\alpha_2^2 \approx 10^{-4}$. As a result, the initial error $\varepsilon$ in 1000 steps was equal to 1,12 $\varepsilon$ , which was quite permissible. This example indicates that the schemes may be used if $\alpha_2^2 \ll 1$ . Later, in case of linear interpolation and $\nu=1$ , the difference equations (3.2.7) will be called "scheme I", and in case of quadratic interpolation when taking additional points and $\nu=1/2$ , they will be called "scheme II ". In both cases we suppose that the stability is verified at least by the model equations and it takes place by fulfilling the CFL condition.

These schemes substantially vary depending on the character of eliminating the calculation errors as well as on the calculation of functions in the region of the unsmooth solution. Scheme I steadily reduces the error, whereas scheme II also reduces the error with respect to the amplitude but its sign may vary. An analogous situation occurs for more complicated equations. It should be noted, that such a behavior can not be regarded as unstable since the stability for the scheme (3.2.8) may be proved strictly.

On the basis of the study of the difference schemes obtained from the Eqs. (3.2.6) the following conclusions may be drawn.

1. The best from the point of view of stability is the scheme of the first order with linear interpolation - scheme I. Apparently, this scheme is monotone and it contains necessary dissipation terms for deriving not very smooth solutions.

2. Scheme II may give rise to annoying fluctuations for unsmooth solutions but it allows us to take wider steps. Therefore, the scheme is useful when finding rather smooth solutions numerically.

3. At $\nu=1/2$ quadratic (or higher) interpolation renders the difference scheme unstable at $\alpha_i = \text{const}$ and stable

only at $\alpha_i^2 = O(\tau)$ . The examination of the equations of hyperbolic type on the basis of many examples showed that the scheme of the first order with respect to $t$ and those of higher order with respect to $x, y$ were stable only at $\alpha_i^2 = O(\tau)$.

## 3. Difference Scheme for Gas Dynamics Equations

Consider steady gas dynamics equations in three dimensions. Take the cylindrical coordinates $z, r, \varphi$ [25] as independent variables and the column vector with components $(p, \eta, \zeta)$ as the unknown vector $u$ where $p$ is the pressure, $\eta = tg\beta$ and $\zeta = tg\gamma$ are the parameters of the velocity vector determined by relations

$$V = V\left\{\cos\beta\cos\gamma, \sin\beta\cos\gamma, \sin\gamma\right\}.$$

Utilize the local Dekartian coordinate system

$$(3.2.17) \qquad k_1 = \frac{V}{|\vec{V}|}, \quad k_2 = \frac{1}{\cos\gamma}\frac{\partial k_1}{\partial\beta}, \quad k_3 = \frac{\partial k_1}{\partial\gamma}$$

related to the velocity vector.

Then the initial system of gas dynamics equations in the coordinate system $k_1, k_2, k_3$, may be written down in the form (3.2.1):

$$u_t + Au_x + Bu_y = f.$$

Here [25] :

$$u_t = \begin{Vmatrix} k_1 \nabla p \\ k_1 \nabla \eta \\ k_1 \nabla \zeta \end{Vmatrix}, \quad u_x = \begin{Vmatrix} k_2 \nabla p \\ k_2 \nabla \eta \\ k_2 \nabla \zeta \end{Vmatrix}, \quad u_y = \begin{Vmatrix} k_3 \nabla p \\ k_3 \nabla \eta \\ k_3 \nabla \zeta \end{Vmatrix},$$

$$f = \frac{1}{r} \begin{Vmatrix} -1/Q\,(M^2-1)\sqrt{(1+\eta^2)}\,\sqrt{(1+\zeta^2)} \\ \zeta^2\sqrt{(1+\eta^2)}/\sqrt{(1+\zeta^2)} \\ -\eta\zeta\,\sqrt{(1+\zeta^2)}/\sqrt{(1+\eta^2)} \end{Vmatrix}, \tag{3.2.18}$$

$$A = \begin{Vmatrix} 0 & 1/Q\,(M^2-1)\sqrt{(1+\eta^2)}\,\sqrt{(1+\zeta^2)} & 0 \\ Q\,(1+\eta^2)\sqrt{(1+\zeta^2)} & 0 & 0 \\ 0 & 0 & 0 \end{Vmatrix},$$

$$B = \begin{Vmatrix} 0 & 0 & 1/Q\,(M^2-1)\,(1+\zeta^2) \\ 0 & 0 & 0 \\ Q\,(1+\zeta^2) & 0 & 0 \end{Vmatrix},$$

## 4. Application of the Difference Scheme for the Calculation of Steady Supersonic Three-Dimensional Flows

By means of the non-steady characteristic network method described above, A. S. Kholodov carried out detailed investigation of the properties of three-dimensional supersonic flow around bodies of different shape(*). Calculations were made by a single algorithm throughout the whole region of disturbance without separating out any singularities beforehand. Bodies of complex form were considered (segments with sharp corners, inverted cones, etc.) in the presence of chemical reactions in the gas and angles of attack up to $\alpha = 25 - 30°$.

Some remarks should be made about the formulation of the problem. The solution is given in the region ABCD (Fig. 36a) bounded by the initially unknown shock wave AB, by the axis of symmetry of the flow AD, by the body CD, and by the ray BC. The boundary conditions are: on the shock wave the Rankine-Hugoniot relations, on AD – the condition of symmetry of the flow, on the body – the conditions of non-penetration.

If the ray BC is completely situated in the supersonic region of the flow, the parameters on it are calculated as within ABCD (the Mach cone is situated within the region). As initial data the shock wave surface was given (paraboloid, for

(*) See also [32] .

example), also a linear distribution of velocity on the body and a linear distribution of parameters between the body and the wave was used.

The region ABCD was covered with a uniform network at the nodes of which the parameters were calculated on layers $t = \text{const}$ until a stationary pattern of flow was attained. For calculating axisymmetrical flows a network was used consisting of 21 points on the body and 11 points across the shock layer (some cases were calculated by means of a 41 x 21 network). The data at angle of attack were obtained essentially by a 11 x 6 x

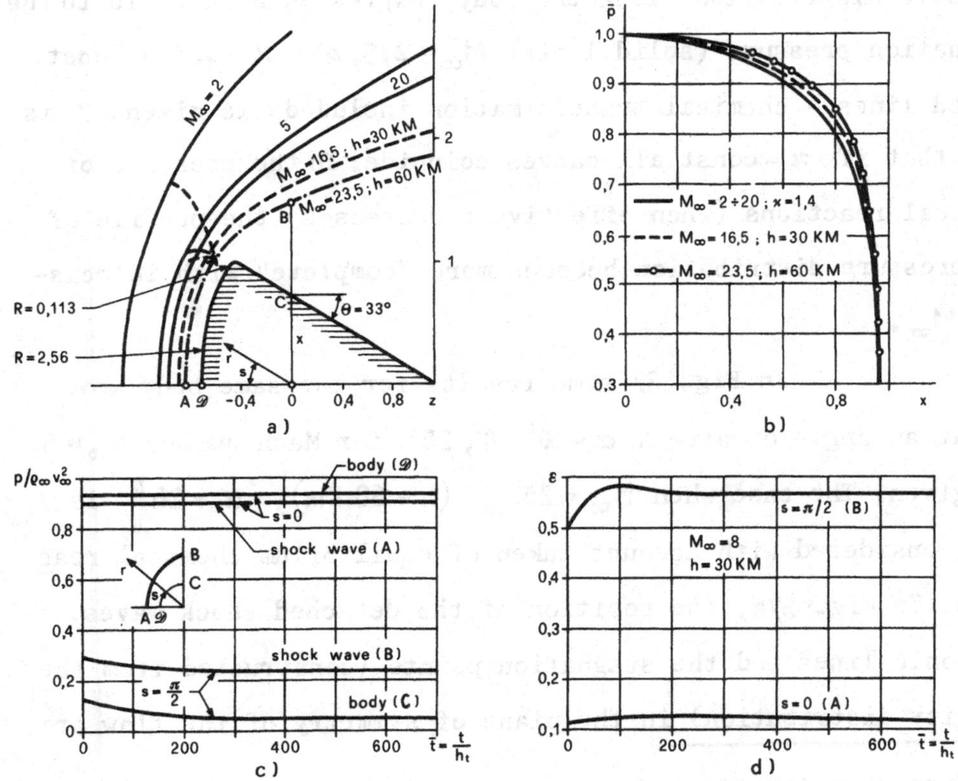


Fig. 36

x 11 network (6 points across the shock layer with quadratic interpolation).

In Fig. 36a,b the flow pattern and surface pressure distribution for supersonic flow of a spherecone body at a zero angle of attack and at various Mach numbers in the undisturbed air stream ( $\varkappa = 1.4$)(*) are given. In Fig. 36a the geometry of the body, the position of detached shock waves and sonic lines with Mach numbers from $M_\infty = 2$ to $M = 23.5$ are represented. In the case $M_\infty = 16.4$ (height $h = 30$ Km) and $M_\infty = 23.5$; ( $h = 60$ Km) chemical reactions were taken into account. In Fig. 36b for the same Mach numbers the pressure distribution along the body, expressed as a ratio to the stagnation pressure (solid lines, $M_\infty = 2, 5, 20$; $\varkappa = 1.4 = \text{const}$, dotted lines - chemical transformation included) is given. It is seen that if $\varkappa = \text{const}$ all curves coincide. In the presence of chemical reactions (when effective $\varkappa$ decreases) the profile of the pressure distribution becomes more "complete" with increasing $M_\infty$.

In Fig. 37 some results for the same body moving at an angle of attack $\alpha = 10°, 20°, 25°$ for Mach number $M_\infty = 6$ are given. The case when $M_\infty = 23.5$ ($h = 60$ Km), $\alpha = 25°$ is also considered with account taken of equilibrium chemical reactions. In Fig. 37a, the position of the detached shock waves, the sonic lines and the stagnation points (constructed from the velocity distribution) in the plane of symmetry of the flow are

(*) Here $\varkappa$-adiabatic index.

given. For these flow conditions there is also given the distribution of pressure (in terms of maximum pressure) on the surface of the body along the nose part $\varphi = 0, \pi$ (Fig. 37b) and in the

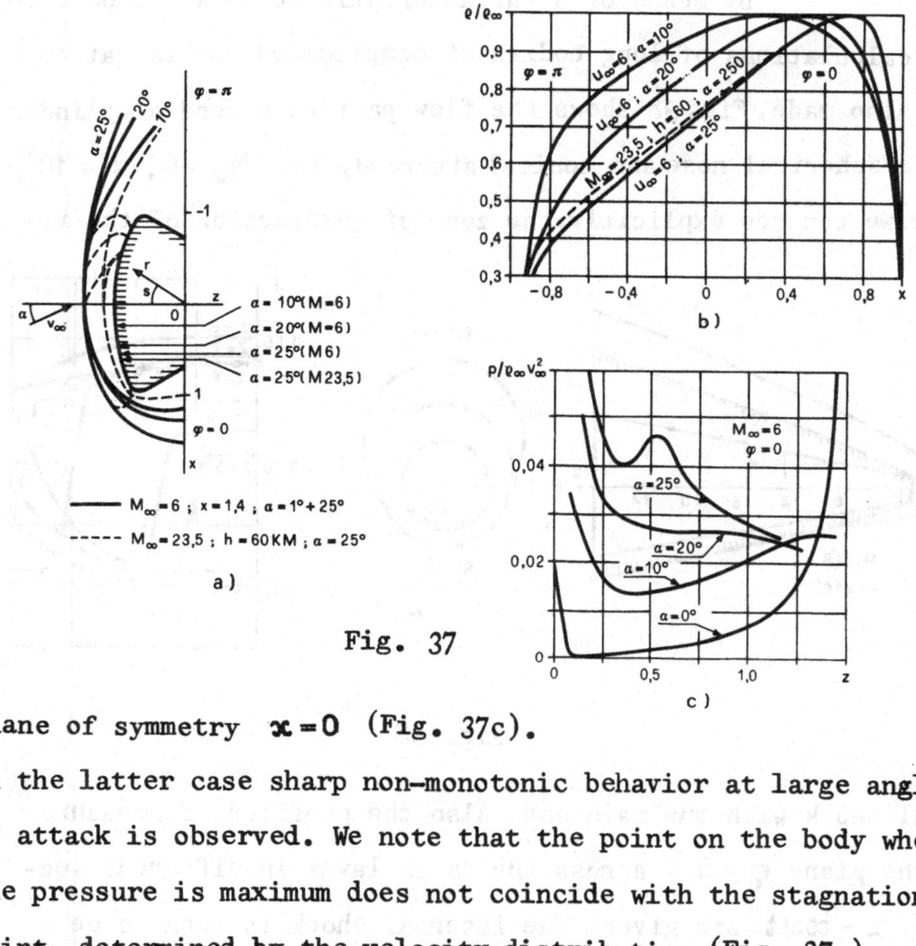


Fig. 37

plane of symmetry $x = 0$ (Fig. 37c).

In the latter case sharp non-monotonic behavior at large angles of attack is observed. We note that the point on the body where the pressure is maximum does not coincide with the stagnation point, determined by the velocity distribution (Fig. 37a).

The approach of parameters to a steady state is shown in Fig. 36c,d. The establishment of steady state pressure (Fig. 36c) for the flow with chemical reactions past a sphere for $M_\infty = 8, h = 30 Km$ is given (at points A, B, C, D). Fig. 36d shows the establishment of the shock wave at points A and B . It is seen that the number of time steps necessary for the attainment of steady state is of the order of 300 – 400.

By means of a characteristic network method complex calculations of long bodies of complicated configuration were also made. Fig. 38 shows the flow pattern around a cylinder with a spherical nose and conical afterbody for $M_\infty = 6$, $\alpha = 10°$, where we can see explicitly the zone of interaction of the in-

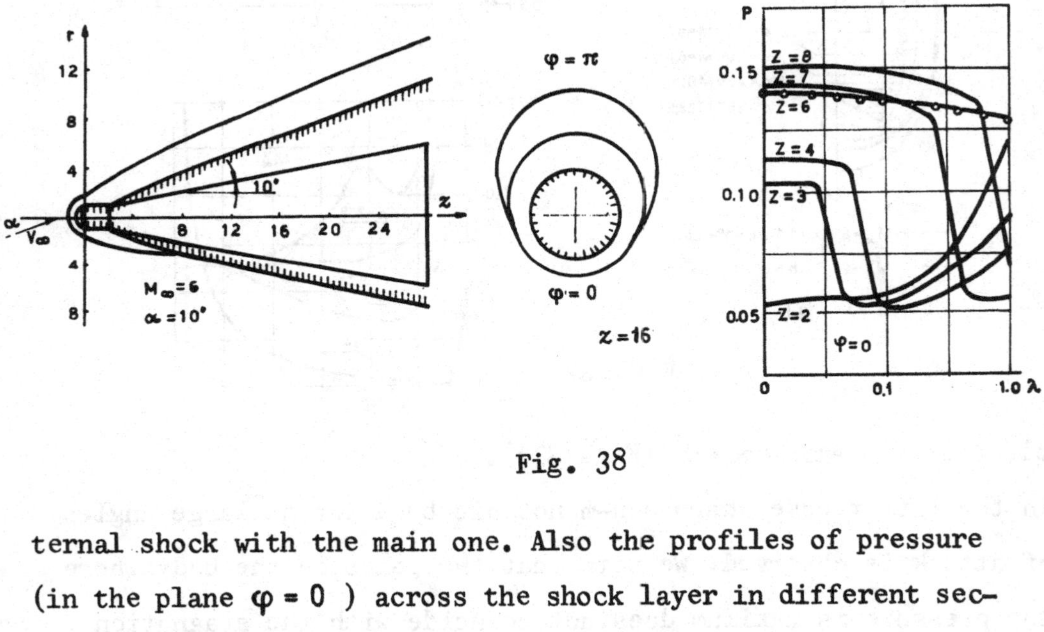


Fig. 38

ternal shock with the main one. Also the profiles of pressure (in the plane $\varphi = 0$ ) across the shock layer in different sections $z = \text{const}$ are given. The internal shock is seen to be "smeared" over 2 – 3 steps with respect to $\lambda$ ( $h_\lambda = 0.05$) though in Fig. 38 it is conventionally plotted as a line . The coordinate $\lambda = (\tau - \tau_0)/(\tau_1 - \tau_0)$, where $\tau = \tau_0(z,\varphi)$ and $\tau = \tau_1(z,\varphi)$ are the equations of the body and nose shock wave, respectively.

## 4. Unsteady "Large Particles" Method for the Solution of Problems of External Aerodynamics

### 4.1. Introduction.

F.H. Harlow (1957) [26] suggested an original "particles in cells" method combining certain advantages of the Langragian and Eulerian approaches. The region of solution here is separated by the fixed (Eulerian) calculation network; however, the continuous medium is interpreted by a discrete model, i.e. the population of "particles" of fixed mass (Langragian network of particles) is considered which move across the Eulerian network of cells. The particles are used for determining the parameters of the fluid itself (its mass, energy, velocity), whereas the Eulerian network is used for determining the parameters of the field (its pressure, density, temperature).

The density in each Eulerian cell is equal to the quotient of division of the total mass of all the particles in a cell by its volume (area), and the law of conservation of mass is always fulfilled automatically. The remaining unknown quantities are found by means of the equations of the laws of conservation of impulse and energy given in the difference form.

The "particles in cells" method allows us to examine complex phenomena of multicomponent media in dynamics; the particles carefully "watch" free surfaces and lines of interfaces between media, interaction of discontinuities and so on. Nevertheless, the discrete method has some drawbacks. The main one underlying the very nature of the method is calculating instability (fluctuations) due to the discrete representation of a continuous medium (the finite number of particles in a cell). It is also rather difficult to get information for considerably rarefied regions in which there are practically no particles left, and so on. The power of modern computers does not permit a considerable increase in the number of particles.

In addition, the calculations by the Harlow method are carried out, essentially, by means of two networks (Eulerian and Langragian), which requires the availability of powerful computers and takes a lot of machine time.

An approximate capacity of machine memory is $(9+3N)AB$ words where $A \times B$ is the dimension of the network, $N$ is the average number of particles in each cell. In practical calculations we use networks consisting of thousands of cells and tens of thousands of particles.

In Diachenko works (1965) [33] Langragian coordinates without fixed network of cells are used for calculating the parameters in "particles". Here one uses "particles" situated at the moment close to each other in space, unlike the previous method in which for this purpose in the finite-difference schemes "particles" with close Langragian coordinates were used. Thus, he essentially rejects any regular calculation network in this approach. An important question here is finding an adequate way of selecting "neighboring Particles".

For gas dynamic problems in the uniform flow it seems more reasonable to use the concept of continuity by considering instead of particles the mass flow across the boundary of Eulerian cells [27, 28, 32, 34] . The density of the flow will be found from the law of conservation of mass written down for the particular cell ("large particle"). It is natural, of course, to make use of the advantageous aspects of the Harlow method (Eulerian-Langragian approach) and of the very process of organizing calculations.

In this paper the numerical schemes of the "large particles" method (O.M. Belotserkovskii, Yu.M. Davidov, 1968) develop exactly according to the above principle [28,32,34] are examined in detail. The main task pursued in this paper is a modification of the "large particles" method in accordance with the problems of external aerodynamics, the elaboration of steady versions of the scheme and its stages, the execution of the method on a computer, carrying out calculations of the flow around vehicles under concrete flight conditions, etc.

In developing and putting this method into practice we tried to utilize the well-known apparatus developed previously for solving problems of external aerodynamics [22,32] retain ing the main principles and approaches while approximating equations with respect to space variables; the divergent form of writing down the initial system was used, as well as the same structure of the calculation networks, and so on.

At the second stage of calculations by this method, the most crucial from the point of view of stability, the construction of diverse difference schemes (by the method of integral relations as well [22]) providing steady calculations in a wide class of problems was performed. The modifications of the first and second stages were also examined (asymmetric difference schemes, the introduction of artificial viscosity of various kinds, additional calculations of density etc.) [28,32,34] .

Single numerical algorithms are considered later. They allow us to carry out calculations for plane and axisymmetric flows without isolating singularities, i.e. both in the regions of smooth flow and in flows with discontinuities (uniform schemes of the "sweeping through" method). The construction of such algorithms results from the uniform description of the hydrodynamic flow. It becomes possible by considering integral laws of conservation or by introducing dissipation terms into gas dynamics equations of an ideal fluid. As a result, the numerical scheme will have approximated viscosity.

Let us outline our approach on the basis of a book B.L. Rodzedestvenski, N.N. Yananko(*).

(*) B.L. Rodzedestvenski, N.N. Yanenko, systems of quasilinear equations and their applications to gas dynamics, Nauka ,1968 Moscow.

The generalized solution is expressed as the limit to the classical solution to some system of quasilinear parabolic equations with small parameters for senior derivatives. If

$$\frac{\partial u}{\partial t} + \frac{\partial \varphi(u)}{\partial x} = f(u) \tag{4.1.1}$$

is the initial system of gas dynamics equations written down in the form of laws of conservation, the corresponding parabolic system has the form

$$\frac{\partial u}{\partial t} + \frac{\partial \varphi(u)}{\partial x} = f(u) + \frac{\partial}{\partial x}\left(\mu B(u)\frac{\partial u}{\partial x}\right), \tag{4.1.2}$$

Here $u = u(x,t)$ is the vector function describing the flow; $f(u), \varphi(u)$ are the vector functions with respect to the vector argument $u$; $B(u)$ is the square matrix; $\mu$ is the small parameter.

The matrix $B(u)$ must be chosen so that the solution $u(x,t)$ to the parabolic system may have sufficient smoothness and at $\mu \to 0$ approach, to some extent, the solution to the hyperbolic system.

In this paper an attempt is also made to calculate by the "large particles" method transonic rotational axisymmetric and plane problems of external aerodynamics for bodies of complex form. As an example, subsonic, transonic and supersonic flows around a cylindrical flat nosed body and a plane step are considered. It should be remarked, that even for supersonic flow conditions it is rather difficult by ordinary methods to calculate the above form of the body due to the presence of a singular corner point, formation of secondary shock waves, very blunt bodies and the large extension of the ("long") body. It is possible, that the above approach will be sufficiently effective for bodies of more complex form as well, for studying the flow with a discontinuity around a stern, characteristics of viscous and separation flows in the trace, and so on.

The work in this direction has been done in the Computing Center of the Academy of Sciences of the U.S.S.R. since 1966. The first results were reported at the 3-d All-Union Congress devoted to theoretical and applied mechanics (Moscow, January, 1968).

## 4.2. Description of the Method

### 1. The initial system of equations.

Consider the motion of an ideal compressible gas. As input equations, we take Eulerian differential equations in the divergent form (equations of continuity, impulse and energy)

$$(4.2.1)\qquad \begin{cases} \dfrac{\partial \rho}{\partial t} + \operatorname{div}(\rho \vec{w}) = 0, \\ \dfrac{\partial \rho u}{\partial t} + \operatorname{div}(\rho u \vec{w}) + \dfrac{\partial p}{\partial x} = 0, \\ \dfrac{\partial \rho v}{\partial t} + \operatorname{div}(\rho v \vec{w}) + \dfrac{\partial p}{\partial y} = 0, \\ \dfrac{\partial \rho E}{\partial t} + \operatorname{div}(\rho E \vec{w}) + \operatorname{div}(p \vec{w}) = 0. \end{cases}$$

It is shown, that in the "large particles" method in place of Eq. (4.2.1) we may use the system of gas dynamics equations written down as laws of conservation in the integral form. It is important, that the difference scheme approximating the initial system of equation is uniform, which enables us to calculate by the "sweeping through" method without separating out singularities.

The form of the system (4.2.1) is similar both for dimensional and dimensionless quantities. Later, we shall make use of the latter by taking the parameters of the oncoming flow as characteristic quantities.

The values of density $\rho$ and velocity $\vec{w}(u, v$ are the components along $x$, $y$, respectively) will be referred to the corresponding values $\rho_\infty, w_\infty$ in the oncoming flow; the pressure $p$ to $\rho_\infty, w_\infty^2$; the specific complete energy $E$ to $w_\infty^2$; linear quantities to the characteristic dimension of the body $R$ (for example, to the radius of the cylinder); the time to $R/w_\infty$.

For closing the system (4.2.1) the equation of state is used

$$p = p(\rho, E, w). \qquad (4.2.2)$$

## 2. Method of splitting.

Consider all the stages of the calculating process separately.

The region of integration is covered by the fixed (Fulerian) network with rectangular cells of sides $\Delta x$ and $\Delta y$. The lues of integers $i$ (along $x$) and $j$ (along $y$) denote the center a cell. The method to be described below is, to some extent, in rmediate between the "particles in cells" method and ordinary inite-difference approaches, since the calculating process is carr..ed out according to the Harlow method but at the same time at all the stages we keep to the structure of the finite-difference scheme.

The steady solution to a problem is the result of stabilization, consequently the entire calculating process consists of multiple repetition of time steps.

The calculation of each step, is turn, is splitted (as it is done by Harlow) into three stages:

I. The "Eulerian stage" at which all the effects related to the motion of fluid are neglected (there is no mass flow across the boundaries of the cells); intermediate values of the unknown parameters of the flow $\tilde{\varphi}(\tilde{u}, \tilde{V}, \tilde{E})$ are given by the fixed Eulerian network;

II. The "Langragian stage" at which the density of the mass flow, with the fluid moving across the boundaries of the cells, is calculated;

III. The final stage consists in determining the final values of the parameters of the flow. $\varphi(u, v, E, \rho)$ for each cell according to the laws of conservation of impulse, energy and mass.

I. Eulerian Stage

At this stage of calculations only the quantities related to the entire cell change but the fluid is assumed to be instantly decelerated. Therefore, the convection terms of the form $\operatorname{div}(\varphi\rho\vec{W}),\ \varphi=\begin{pmatrix}1\\u\\v\\E\end{pmatrix}$, corresponding to the effects of displacement are eliminated from the Eqs. (4.2.1).

It follows, in particular, from the equation of continuities field will be "frozen", consequently, in the remaining equations (4.2.1) we shall be able to take out $\rho$ from the differential sign and solve the Eqs. (4.2.1) in terms of the time derivatives with respect to $u, v, E$.

Then

$$\left.\begin{aligned} \rho\frac{\partial u}{\partial t}+\frac{\partial p}{\partial x}&=0\,,\\ \rho\frac{\partial v}{\partial t}+\frac{\partial p}{\partial y}&=0\,,\\ \rho\frac{\partial E}{\partial t}+\operatorname{div}(p\vec{W})&=0\,. \end{aligned}\right\} \qquad (4.2.3)$$

For improving the stability of calculations at the first stage in the Eqs.(4.2.3) $p$ is often replaced by $p = p+q$ . Here $q$ is the artificial viscous pressure, for example, of Landshoff type [27]

$$q = \begin{cases} -BC\rho h \dfrac{\partial u}{\partial s} & \text{for } \dfrac{\partial u}{\partial s} < 0 , \\ 0 & \text{for } \dfrac{\partial u}{\partial s} \geq 0 , \end{cases} \tag{4.2.4}$$

where $B$ is the constant coefficient, $C$ is the same velocity, $h$ is the dimension of the cell in the direction $s$ ($x$ or $y$).

Thus, the viscosity introduced becomes apparent only on compressible waves, on shocks, and it is missing on rarefied waves (where the coefficent of viscosity "vanishes"), which increases the accuracy of difference calculations with artificial viscosity. For $q = 0$ the scheme of the first stage is unstable.

Write down the Eqs. (4.2.3) in the form of finite-difference equations of the first order of accuracy with respect to time and space

$$\begin{aligned} \tilde{u}^n_{i,j} &= u^n_{i,j} - \frac{p^n_{i+1/2,j} - p^n_{i-1/2,j}}{\Delta x} \frac{\Delta t}{\rho^n_{i,j}} , \\ \tilde{v}^n_{i,j} &= v^n_{i,j} - \frac{p^n_{i,j+1/2} - p^n_{i,j-1/2}}{\Delta y} \frac{\Delta t}{\rho^n_{i,j}} , \\ \tilde{E}^n_{i,j} &= E^n_{i,j} - \left[ \frac{p^n_{i+1/2,j} u_{i+1/2,j} - p_{i-1/2,j} u_{i-1/2,j}}{\Delta x} + \right. \\ &\quad \left. + \frac{p^n_{i,j+1/2} v_{i,j+1/2} - p^n_{i,j-1/2} v_{i,j-1/2}}{\Delta y} \right] \frac{\Delta t}{\rho^n_{i,j}} . \end{aligned} \tag{4.2.5}$$

Here the quantities with fractional indices refer to the boundaries of the cells, for example, $p^n_{i+1/2,j} = \frac{p^n_{i,j} + p^n_{i+1,j}}{2} + q_{i+1/2,j}$ and so on ; $\tilde{u}, \tilde{v}, \tilde{E}$ are the intermediate values of the flow parameters obtained by assuming the field $\rho$ "frozen" on the layer $t^n + \Delta t$ .

## II Langragian Stage

At this stage for $t^n + \Delta t$ we find mass flows $\Delta M^n$ across the boundaries of the cells. In addition, we assume that the whole mass is carried over only by the component of velocity normal to the boundary. Thus, for example,

$$\Delta M^n_{i+1/2,j} = < \rho^n_{i+1/2,j} > < u^n_{i+1/2,j} > \Delta y \Delta t \,. \qquad (4.2.6)$$

The sign $< \; >$ denotes the values of $\rho$ and $u$ on the boundaries of a cell. The selection of these quantities is of great importance, since it considerably affects the stability and accuracy of calculations. Taking into account the direction of the flow across the particular boundary is characteristic of all the forms of writing down $\Delta M^n$.

a) $\Delta M^n$ is expressed in terms of the formulas of the second order of accuracy [27,28,34] . For example, is calculated in the following way. If $\tilde{u}^n_{ij} + \tilde{u}_{i+1,j} > 0$ and

$$\tilde{u}^n_{i+\frac{1}{2},j} = \tilde{u}^n_{i,j} + \left(\frac{\partial \tilde{u}}{\partial x}\right)^n_{i,j} \frac{\Delta x}{2} = \tilde{u}^n_{i,j} + \frac{\tilde{u}^n_{i+1,j} - \tilde{u}_{i-1,j}}{4} > 0$$

then

$$\Delta M^n_{i+\frac{1}{2},j} = \left(\tilde{u}^n_{ij} + \frac{\tilde{u}^n_{i+1,j} + \tilde{u}_{i-1,j}}{4}\right)\left(\varrho^n_{i,j} + \frac{\varrho^n_{i+1,j} - \varrho^n_{i-1,j}}{4}\right)\Delta y\, \Delta t \tag{4.2.7}$$

etc.

b) $\Delta M^n$ is given in terms of the formulas of the first order of accuracy [28, 34, 35]

$$\Delta M^n_{i+\frac{1}{2},j} = \begin{cases} \varrho^n_{ij} \dfrac{\tilde{u}^n_{ij} + \tilde{u}^n_{i+1,j}}{2} \Delta y\, \Delta t & \text{if} \quad \tilde{u}^n_{ij} + \tilde{u}^n_{i+1,j} > 0 \\ \varrho_{i+1,j} \dfrac{\tilde{u}^n_{ij} + \tilde{u}^n_{i+1,j}}{2} \Delta y\, \Delta t & \text{if} \quad \tilde{u}^n_{ij} + \tilde{u}^n_{i+1,j} < 0 \end{cases} \tag{4.2.8}$$

etc.

Here it is possible, without losing stability, to utilize the values of velocity on the preceding time layer $t^n$ ; consequently, in the particular case we can, at first, carry out calculations at the Lagrangian stage and then at the Eulerian one.

## III. Final Stage.

At this stage we find final fields of Eulerian flow parameters at time $t^{n+1} = t^n + \Delta t$ . As was remarked, the equations at this stage represent the laws of conservation of mass $M$, impulse $\vec{P}$ and full energy $E$ written down for the particular cell in the difference form

$$M^{n+1} = M^n + \left(\Sigma \, \Delta \, M^n\right) \quad \text{boundary}, \tag{4.2.9}$$

$$\vec{P}^{\,n+1} = \vec{P}^n + \left(\Sigma \, \Delta \, \vec{P}^n\right) \quad \text{boundary}, \tag{4.2.10}$$

$$E^{n+1} = E + \left(\Sigma \, \Delta \, E^n\right) \quad \text{boundary}, \tag{4.2.11}$$

These equations state that inside the flow field there are no sources and sinks $M$ , $p$ and $E$;their variation during time $\Delta t$ is only due to interaction on the external boundary of the flow region.

For this reason, the final values of the flow parameters $\rho, x = (u, v, E)$ on the adjacent layer are calculated by the formulas

$$\rho^{n+1}_{i,j} = \rho^n_{i,j} + \frac{\Delta M^n_{i-1/2,j} + \Delta M^n_{i,j-1/2} - \Delta M^n_{i,j+1/2} - \Delta M^n_{i+1/2,j}}{\Delta x \, \Delta y}, \tag{4.2.12}$$

$$x^{n+1}_{i,j} = \tilde{x}^n_{i,j} + \frac{\tilde{x}^n_{i-1,j} \Delta M^n_{i-1/2,j} + \tilde{x}^n_{i,j-1} \Delta M^n_{i,j-1/2} - \tilde{x}^n_{i+1,j} \Delta M^n_{i+1/2,j} - \tilde{x}^n_{i,j+1} \Delta M^n_{i,j+1/2}}{\rho^{n+1}_{i,j} \Delta x \, \Delta y}$$

It is easy to show that Eqs. (4.2.12)-(4.2.13) satisfy Eqs.(4.2.9)-(4.2.11)[28] .

General variation of energy and impulse during time $\Delta t$ is equal to the sum of these variations at Eulerian and at final stages. Therefore, inside the flow region a rigorous conservation of quantities $E$ and $\vec{p}$ take place.

Thus, we showed that at the final stage the laws of conservation of mass, impulse and energy were fulfilled (the difference scheme was conservative) [28, 34] .

In practical calculation the formula (4.2.13) was altered for automatic determination of the direction of the flow. Introduce function $D_{i,j}$ and index all the sides of the cell $(i,j)$ by $k = 1,2,3$ and $4$, as is shown below

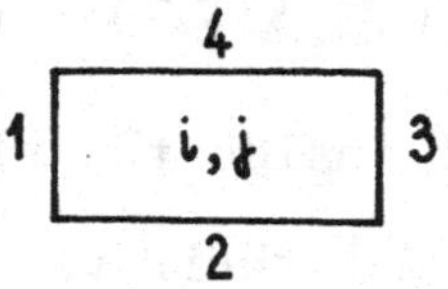


Express the values of $D_{i,j}(k)$ related to the side $k$ as follows

(4.2.14) $$D_{i,j}(k) = \begin{cases} 1, & \text{if the fluid flows into the cell } (i,j) \text{ through the side } k. \\ 0, & \text{If the fluid flows out of the cell } (i,j) \text{ through the side } k. \end{cases}$$

Then the final values of the flow parameters $x(u, v, E)$ on the new time layer $t^{n+1} = t^n + \Delta t$ are given as follows [27]

(4.2.15)
$$x_{i,j}^{n+1} = \Big( D_{i,j}(1)\tilde{x}_{i-1,j}^{n} \Delta M_{i-1/2,j}^{n} + D_{i,j}(2)\tilde{x}_{i,j-1}^{n} \Delta M_{i,j-1/2}^{n} + $$
$$+ D_{i,j}(3)\tilde{x}_{i+1,j}^{n} \Delta M_{i+1/2,j}^{n} + D_{i,j}(4)\tilde{x}_{i,j+1}^{n} \Delta M_{i,j+1/2}^{n} + $$
$$+ \tilde{x}_{i,j}^{n} \Big\{ \rho_{i,j}^{n} \Delta x \Delta y - \big[1 - D_{i,j}(1)\big] \Delta M_{i-1/2,j}^{n} - \big[1 - D_{i,j}(2)\big] \Delta M_{i,j-1/2}^{n} - $$
$$- \big[1 - D_{i,j}(3)\big] \Delta M_{i+1/2,j}^{n} - \big[1 - D_{i,j}(4)\big] \Delta M_{i,j+1/2}^{n} \Big\} \Big) \Big/ \rho_{i,j}^{n+1} \Delta x \Delta y .$$

The calculating process is accomplished. Draw attention to the fact that we use the equation for full energy E , which provides conservativeness, i.e. complete divergence of the difference scheme.

3. Boundary Conditions

The region in which the calculations were carried out, as well as the calculation network are shown in Fig. 39. The parameters of the undisturbed flow were used as initial data. The boundary conditions were taken in the following way: on the left boundary (AB)- the conditions of the oncoming flow of a gas were used; on the axis of symmetry (OA) and on the body (OKD)-the conditions of symmetry and ordinary conditions on the wall (conditions of nonpenetration); on the upper and right boundaries of the region (BCD)- extrapolation of the flow parameters outside the region considered was done.

In order not to violate the uniformity of calculations and not to use special formulas for boundary cells, layers of imaginary cells are introduced along all the boundaries where the parameters of the adjacent cells of the flow are placed. The number of such layers is determined by the order of the difference scheme (for the first order- one layer). In this condition, it is necessary to distinguish 2 kinds of boundaries: a rigid boundary (or an axis of symmetry) and an open boundary of the calculated region.

In the first case, the component of velocity normal to the boundary changes its sign and the remaining flow parameters are taken unchanged. Thus, on walls the normal component of velocity vanishes and in this way the condition of nonpenetration is realized. It will be shown below, that another type of boundary conditions is possible: a wall devoid of sliding. In this case, both components of velocity change their sign and the whole velocity vector vanishes on the wall (the condition of sticking).

Across the open boundaries of the region, fluid may flow in or out and some conditions of continuous motion must be satisfied. Let fluid flow into the rectangular network from the left. Then the parameters of the oncoming flow are given here. On the remaining open boundaries of the region, extrapolation of the flow parameters is performed "outwards", i.e. the values of the parameters from the nearest (to the boundary) layer are carried out into the imaginary layer (extrapolation of zero order). A more complicated formulation of conditions and more exact extrapolation (linear, quadratic) are possible as well.

In the supersonic flow the character of boundary conditions (extrapolation outwards) does not entail any complications, since the disturbances are "carried away" by the flow. In the subsonic flow the situation is more complicated since the influence of the boundary must be made negligible here.

It is obvious, that the external boundary of the region must be situated appreciably far from sources of disturbances, then methods of extrapolation of the flow "into outside" are possible. This question will be discussed in detail later. Note only, that the main principle of formulating conditions lies in the fact that across open boundaries of the region any appreciable disturbances must not penetrate into the calculated region.

## 4.3 Viscous Effects

As was noted, in the approach considered the uniform difference schemes are used. They allow us by single algorithms to calculate both the regions of smooth flow and flow with a discontinuity by means of the "sweeping through" method. It is achieved by using finite-difference schemes with approximated viscosity. Discuss briefly this question.

Gas dynamics equations for an inviscid gas are taken as input equations; however, viscous effects are inherent in our difference scheme as well. First, they are due to the introduction into the scheme an explicit term with artificial viscosity ("viscous pressure") $q$ and, second, to the presence of intrinsic network viscosity dependent on the structure of finite-difference equations.

The form of approximated viscosity and the evaluation of stability of the scheme may be attained by expanding difference equations at all three stages into Taylor series. Then terms of zero (lowest) order must represent the input difference equations, the account taken of terms of higher order (errors due to expansion) in the expansion allows us to define the structure of approximated viscosity.

Write out, for the sake of simplicity in the one-dimensional case, the first differential approximation of our difference scheme. Express, for example, $u^n_{i+1}$ as function $u(x+\Delta x, t)$ and expand each term of finite- difference equations into Taylor series in the vicinity of the point $(x,t)$.

Then, by calculating $\Delta M$ by means of the formulas (4.2.7), we obtain

$$\frac{\partial \rho}{\partial t} + \frac{\partial \rho u}{\partial x} = 0$$

(4.3.1)
$$\frac{\partial \rho u}{\partial t} + \frac{\partial (p+\rho u^2)}{\partial x} = -\frac{\partial q}{\partial x} + \frac{\partial}{\partial x}\left(\rho\varepsilon\frac{\partial u}{\partial x}\right),$$

$$\frac{\partial \rho E}{\partial t} + \frac{\partial}{\partial x}\left[u(p+\rho E)\right] = -\frac{\partial q u}{\partial x} + \frac{\partial}{\partial x}\left(\rho\varepsilon\frac{\partial E}{\partial x}\right).$$

Using the formulas (4.2.8), we have

(4.3.2a)
$$\frac{\partial \rho}{\partial t} + \frac{\partial \rho u}{\partial x} = \frac{\partial}{\partial x}\left(\varepsilon\frac{\partial \rho}{\partial x}\right),$$

$$\frac{\partial \rho u}{\partial t} + \frac{\partial (p+\rho u^2)}{\partial x} = - \frac{\partial q}{\partial x} + \frac{\partial}{\partial x}\left(\varepsilon \frac{\partial \rho u}{\partial x}\right) + \rho\varepsilon \frac{\partial^2 u}{\partial x^2} ,$$

(4.3.2b)

$$\frac{\partial \rho E}{\partial t} + \frac{\partial}{\partial x}\left[u(p+\rho E)\right] = - \frac{\partial q u}{\partial x} + \varepsilon \frac{\partial^2(\rho E)}{\partial x^2} + \frac{\partial \varepsilon}{\partial x}\frac{\partial \rho E}{\partial x} ,$$

where $\varepsilon = \frac{1}{2}|u|\Delta x$.

Differential approximations for two-dimensional problems are written out in a similar way.

On the left-hand side in eqs. (4.3.1) and (4.3.2) we have obtained the exact expressions for input differential equations and the terms which are due to the presence of "viscous" effects in the difference equations are written out on the right-hand side. Terms with $q$ are due to the apparent introduction of artificial viscosity and terms with $\varepsilon$ are determined by network viscosity arising when we replace exact differential equations by finite-difference ones.

It is easy to see, that with mesh refinement $(\Delta x \to 0)\, \varepsilon \to 0$ equations of differential approximation change into the exact system of input equations as well. In concrete calculations (due to finiteness of $\Delta x, \Delta t \dots$) terms containing $\varepsilon$ are always implicitly present in the difference schemes. They, in turn, are analogous to dissipation terms of the Navier-Stokes equations. In addition, the coefficient of network viscosity $\varepsilon$ dependent on local velocity of the flow and the dimension of the differnce network plays the role of the coefficient of real viscosity $\nu$ .

In the two-dimensional case for (4.2.7), it follows from equations of impulse that network viscosity (at $q=0$) has the tensor form

$$\Delta = \frac{1}{2}\rho \begin{Vmatrix} u\Delta x\dfrac{\partial u}{\partial x} & v\Delta y\dfrac{\partial u}{\partial y} \\ u\Delta x\dfrac{\partial v}{\partial x} & v\Delta y\dfrac{\partial v}{\partial y} \end{Vmatrix} = \frac{1}{2}\rho\cdot\vec{u}\cdot\Delta r\nabla\vec{u}, \tag{4.3.3}$$

where $\Delta\vec{r} = \Delta x\cdot\vec{i} + \Delta y\cdot\vec{j}$ .

It is seen, that network viscosity due to the presence of the fixed vectors $\vec{\Delta r}$ and $\vec{u}$ is not invariant with respect to Galileo transformations and practically it becomes appreciable only in zones with high gradients: on the shock wave, near the surface of the body, in the flow with a discontinuity and so on. For this reason, the coefficient of network viscosity (and, therefore, the width of the "smeared" shock wave obtained as well) depend on the value of the local flow velocity. In regions of smooth flow where gradients of flow parameters are small the influence of network viscosity is negligible.

As will be shown below, in some cases (the formulas (4.2.8) were used to calculate $\Delta M''$) network viscosity provides stable calculations without introducing the explicit term with pseudoviscosity; in using the formulas (4.2.7), in regions where the value of local velocity as compared to the sonic velocity is negligible, the introduction of a term with $q$ is required for obtaining a stable solution.

As has been noted, the presence of approximated viscosity in the finite-difference scheme provides parabolic properties to the solution obtained, which enables us by single algorithms to carry out calculation by the "sweeping-through" method without isolating singularities. Viscous effects result in the "smearing" of shock waves (even at $q = 0$) into several calculation cells; a comparatively wide boundary layer is formed near the body, be hind the corner point under certain conditions the flow with a discontinuity is observed; the vortex zone is formed, and so on.

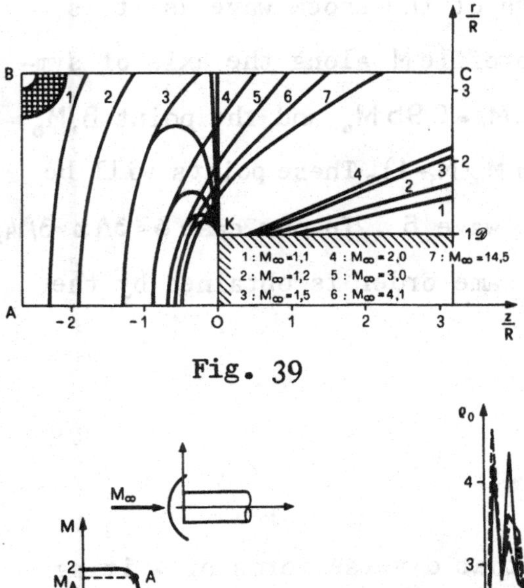


Fig. 39

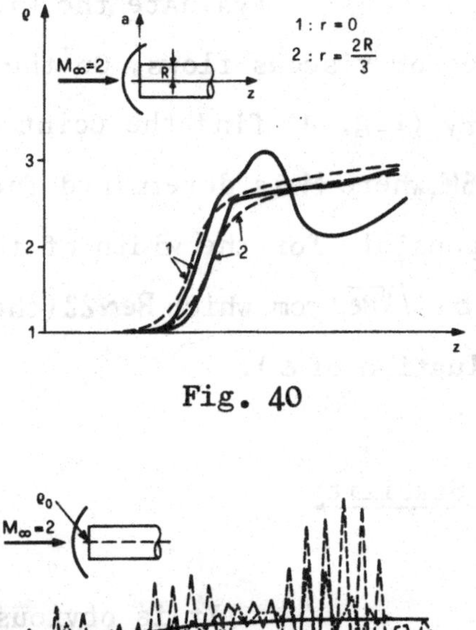


Fig. 40

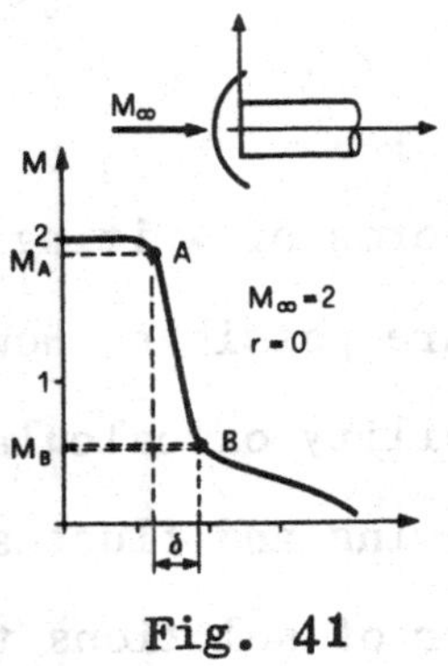


Fig. 41

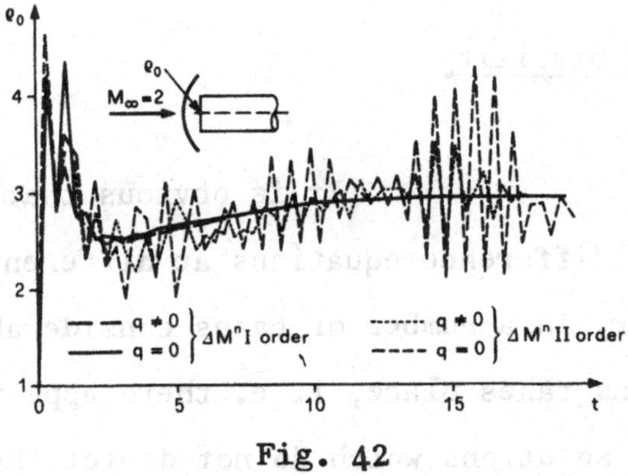


Fig. 42

In Fig. 40 the density profiles $\rho$ on lines $r = \text{const}$ ahead of the cylindrical flat-nose body are given. Solid lines correspond to Eq. (4.3.1) and dots to Eq. (4.3.2). In both cases viscous pressure $q$ with the same coefficient $B$ was used. It is seen, that the presence of the term of network viscosity in the equation of continuity contributes to greater smearing of the shock wave in the second case.

One may evaluate the influence of approximated viscosity by means of the "width" $\delta$ of the shock wave or the value $\varepsilon = \frac{1}{2} u \Delta x$ .

Evaluate the width of the shock wave as it is done for viscous flows. On the profile $M$ along the axis of symmetry (Fig. 41 find the point $A, M_A = 0{,}95\, M_\infty$ and the point $B, M_B = 1{,}05 M_1$ where $M$ is determined form $M_\infty \cdot M_1 = 1$). These points will be responsible for the width of the wave $\delta$ . In our case $\delta \cong 3\Delta x = 3/14$; but $\delta \sim 1/\sqrt{Re}$ from which $Re \sim 22$ (the same order is obtained by the evaluation of $\varepsilon$).

## 4.4 Stability

It is obvious that diverse forms of writing out difference equations at different stages are possible. However, in a number of cases considerable instability of calculations takes place, i. e. there appear fast growing and fluctuating solutions which do not depict the behaviour of solutions to input differential equations.

The above difference schemes are stratified and the difference equations,are essentially,nonlinear with variable coefficients. It makes impossible to use the Fourier method (based upon the investigation of nonlinear equations with constant coefficients) for analyzing stability of the whole system. The Fourier method, essentially, implies linearization of equations in the vicinity of the flow with constant parameters and does not take into account linear effects (the influence of gradients of the flow) which in a number of cases are real sources of instability.

Use, therefore, a heuristic approach for the analysis of stability of difference schemes, suitable for nonlinear equations and based upon the consideration of their differential approximations. *)

Resume now the investigation of our difference scheme.

Consider, for example, what is the contribution to instability of different forms of writing down equations of continuity when assuming equations of impulse and energy to be stable.

---

*) C. W. Hirt. Heuristic Stability Theory for Finite-Difference Equations, Jour. of Com. Ph., 1968, v. 2, N4, 339-355.

If $\Delta M^n$ is calculated by formulas of the second order of accuracy (4.2.7), we obtain by expanding the given difference equations into Taylor series and retaining terms with $\partial^2\rho/\partial x^2$

$$\frac{\partial \rho}{\partial t} + \frac{\partial \rho u}{\partial x} = \Delta_1 - \frac{\Delta t}{2}\left(u^2 + c^2\right)\frac{\partial \rho}{\partial x^2} . \tag{4.4.1}$$

In the case of calculating $\Delta M^n$ by means of the formula of the first order accuracy (4.2.8), we have

$$\frac{\partial \rho}{\partial t} + \frac{\partial \rho u}{\partial x} = \Delta_1^* + \left\{\frac{\Delta x}{2}|u| - \frac{\Delta t}{2}\left(u^2+c^2\right) - \frac{\Delta x^2}{4}\frac{\partial u}{\partial x}\right\}\frac{\partial^2 \rho}{\partial x^2} \tag{4.4.2}$$

$\Delta_1, \Delta_1^*$ are terms of the first differential approximation proportional to $\Delta x$. We have

$$\Delta x \cong 0{,}071 \,, \quad \Delta t \cong 0{,}0071 \,, \quad \rho_{-\infty} = 1 \,, \quad u_{-\infty} = 1 .$$

In practical calculations, when shock waves tangent discontinuities and rarefied waves appears, we have

$$\rho|u| = 1 \,, \quad \left|\frac{\partial u}{\partial x}\right|\delta x < 0{,}3 \,, \quad \left|\frac{\partial \rho}{\partial x}\right|\delta x < 2 .$$

Thus we see, that in Eq. (4.4.2) the coefficient at $\frac{\partial^2 \rho}{\partial x^2}$ is positive and in Eq. (4.4.1) negative, i.e. the scheme (4.4.1) is unstable and the scheme (4.4.2) is stable. In Fig. 40 the profiles $\rho$ for these two cases are given. It is seen, that for Eq. (4.4.1) substantial instability originates immediately behind the shock wave.

In the first case, stability of calculation is attained by introducing into the scheme the explicit term of viscous pressure $q$. It is possible to show [28] that using for $\Delta M^n$ the formula (4.2.8) renders our whole difference scheme stable even without introducing the term with $q$ . Expand the difference equations at all three stages into Taylor series to within terms of the first order in time and second-order terms with respect to space, inclusively. It appears, that coefficients at $\partial^2\varphi/\partial x^2$ and $\partial^2\varphi/\partial y^2$ $(\varphi=\{\rho,u,v,E\})$ have the analogous form, consequently restrict only to the consideration of one-dimensional equations of motion.

Note, that in the above method, difference equations at Eulerian stage (symmetric differences) are unstable. It may be shown, that the introduction of viscous pressure $q = -BC\rho\,\Delta x\,\frac{\partial u}{\partial x}$ , where $B$ is the empirical constant, $c$ is the sonic velocity renders the Eulerian stage stable.

Examine stability by the Fourier method in the approximation of "frozen" coefficients. In this approximation equations for the variation of the solution coincide with the input equation.

It is seen, that in absence of viscous pressure the scheme is unstable and the introduction of $q$ renders it stable ($\mathrm{Re}\,\omega\leq 0$). Nevertheless, for the sake of simplicity of writing down equations and convenience of calculations, one may use an automatically unstable (at the Eulerian stage) scheme with $q=0$ . It does not practically affect the final result, since the stability of the whole difference scheme at all three stages is of importance.

All stated above is illustrated by Fig. 42 in

which the character of stabilization of $\rho$ at the stagnation point

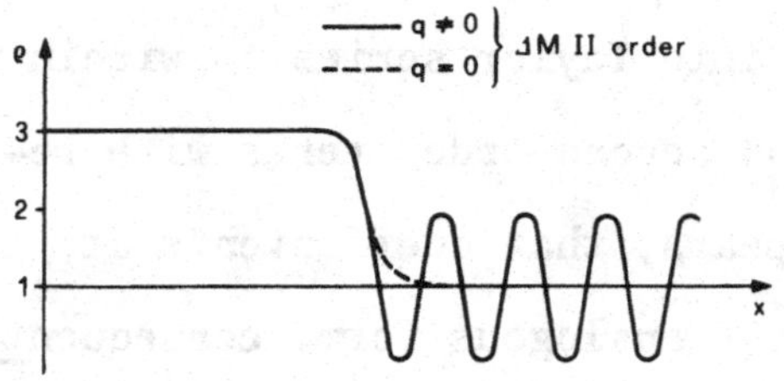


Fig. 43

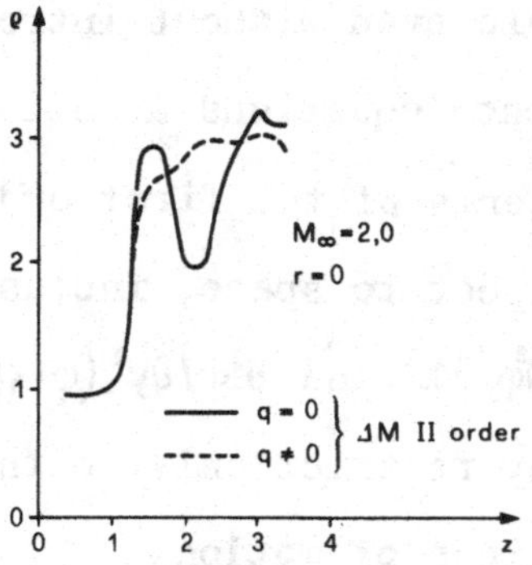


Fig. 44

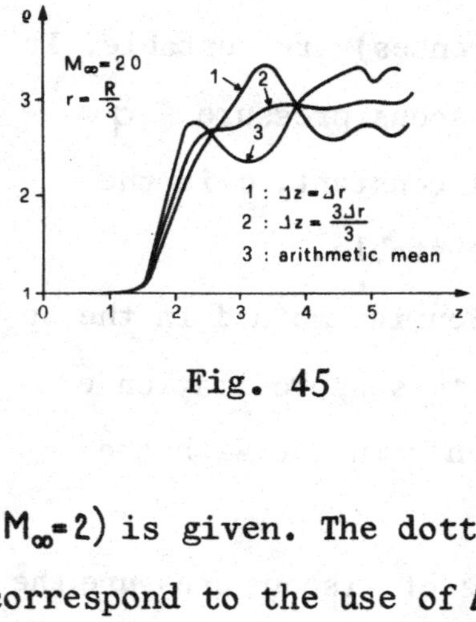


Fig. 45

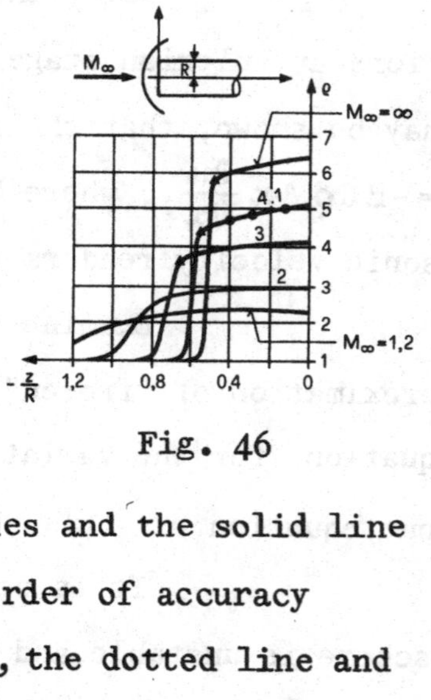


Fig. 46

($M_\infty$=2) is given. The dotted line with primes and the solid line correspond to the use of $\Delta M$ of the first order of accuracy (4.2.8) with q and without q, respectively, the dotted line and primes to the use of $\Delta M$ of the second order of accuracy (4.2.7), with q and without q as well. We see, that in case of $\Delta M$ calculated in terms of Eqs. (4.2.7), instability which may be eliminated only by introducing q develops. Obtaining $\Delta M$ in terms of Eq. (4.2.8) renders the calculation stable in any case (the introduction of q only slightly accelerates convergence). By using

Eq. (4.2.8) stabilization with respect to all the parameters was practically attained to within 0.0001% .

The periodic character of instability inherent in the particular scheme using the formulas (4.2.7), should be noted as well. When approaching the unknown (stable) solution the gradients of the flow parameters in the regions of smooth flow become negligible. The viscous pressure and the viscous terms of the first differential approximation become negligible as well, and in these regions instability begins to develop again. When it develops to the extent, that the gradients become high, the viscous terms begin to damp it again and so on, "auto-oscillations" arise. Such a periodic structure of instability is clearly seen in Fig. 42.

## 4.5. Results of Numerical Calculations

Calculations of subsonic, transonic and supersonic flows around a cylindrical flat-nose body and a plane step were carried out by means of the method of "large particles" (main importance was attached to transonic flows) [28], [31], [36]. The problem was chosen to make it possible to carry out the calculations for a whole spectrum of flow conditions ranging from purely subsonic flow to hypersonic (including transonic regions and transitions through sonic velocity) [28], [34], [36], [37].

The calculations were realized on a BESM- 6 computer. The region of integration was divided into 40 (along the vertical line) and from 20 to 60 (along the horizontal line) calculating cells. Here we did not single out the shocks in the flow beforehand. Such a "sweeping through" method turned out to be reasonable due to the presence of the lines of discontinuity within the disturbed region.

Questions of stability were solved by methods described previously (Fig. 42). Most of the calculations were carried out by means of the scheme utilizing for $\Delta M^n$ formulas of the first order of accuracy (4.2.8). In Fig. 43 the solutions to a problem of the motion of the one-dimensional shock wave obtained with artificial viscosity $q$ (primes) and without it (solid line) are compared (the formulas (4.2.7) were used). The smoothing effect of artificial viscosity is apparent. In Fig. 44 there are given profiles $\rho$ on the axis of symmetry of the flow around the cylindrical flat-nose body obtained for similar cases for $\Delta M^n$ by means of the formulas of the second order of accuracy (4.2.7) with artificial viscosity $q$ (primes) and without $q$ (solid line). The selection of an optimum dimension of the calculation network and the averaging of the parameters on the time layer are also of great importance ("peaks" of instability due to subdividing the step of the network periodically appear in the counterphase). The smoothing effect of the averaging is illustrated in Fig. 45.

Dwell, at first, on the results of calculations

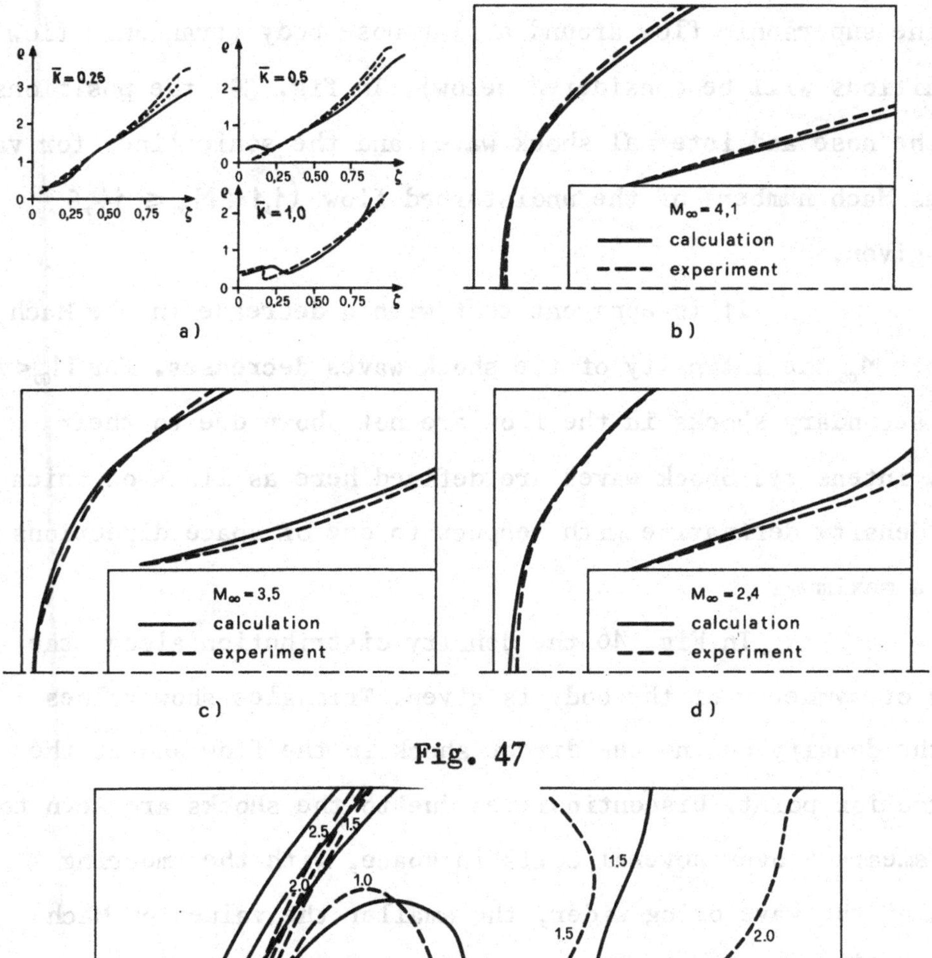
ρ
$\bar{K} = 0,25$
$\bar{K} = 0,5$
$\bar{K} = 1,0$
0
0,25
0,50
0,75
ζ
a)
$M_\infty = 4,1$
calculation
experiment
b)
$M_\infty = 3,5$
calculation
experiment
c)
$M_\infty = 2,4$
calculation
experiment
d)

Fig. 47

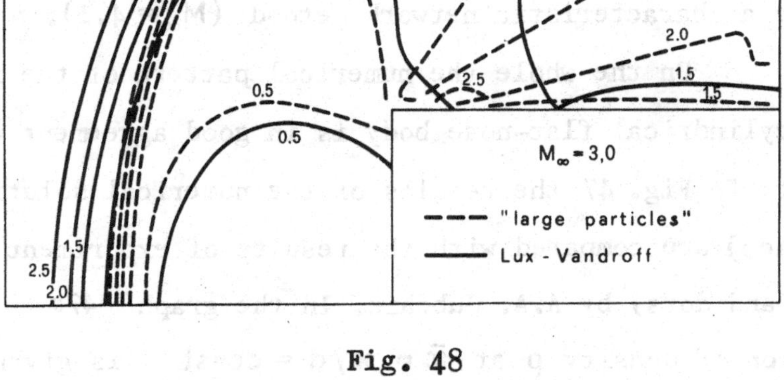
0.5
1.0
1.5
2.0
2.5
$M_\infty = 3,0$
"large particles"
Lux - Vandroff

Fig. 48

of the supersonic flow around a flat-nose body (transonic flow conditions will be considered below). In fig. 39 the positions of the nose and internal shock waves and the sonic lines for various Mach numbers of the undisturbed flow $(1{,}1 \leq M_\infty \leq 14{,}5)$ are given.

It is apparent that with a decrease in the Mach number $M_\infty$ the intensity of the shock waves decreases. For $M_\infty < 2$ the secondary shocks in the flow are not shown due to their weak intensity. Shock waves are defined here as lines on which the density derivative with respect to one of space directions has a maximum.

In Fig. 46 the density distribution along the axis of symmetry of the body is given. Triangles show values of the density behind the direct shock in the flow and at the stagnation point. Discontinuities due to the shocks are seen to be "smeared" over several cells in space, with the smearing band of the wave being wider, the smaller the values of Mach numbers $M_\infty$. Circles in Fig. 46 indicate the results of calculations by a characteristic network method $(M_\infty = 4.1)$.

On the whole the numerical pattern of the flow around a cylindrical flat-nose body is in good agreement with experiment. In Fig. 47 the results of the numerical solution (solid lines) are compared with the results of experiment (dotted lines and dots) by A.A. Gubchik. In the graph 47a the distribution of density $\rho$ at $\bar{x} = x/d = \text{const}$ is given; along the abscissa the coordinate $\zeta = (y-h)/(H-h)$ is plotted, where $h$ and $H$ are the equations of the body and shock wave, respectively. The results are in fairly good agreement, especially in the flow field and on the shock wave.

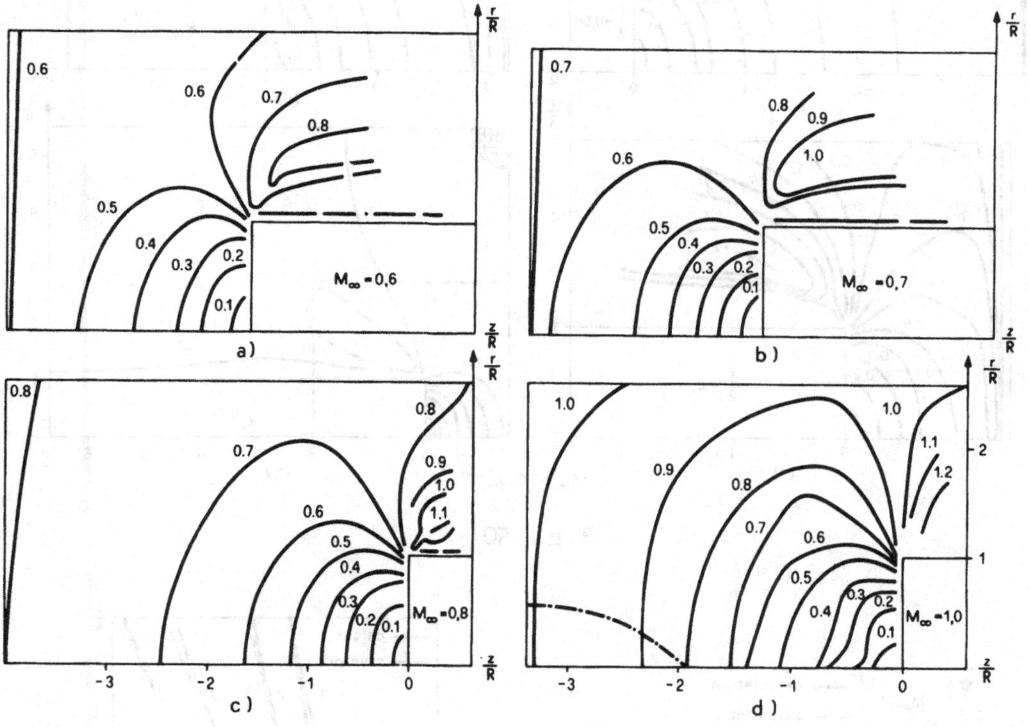

Fig. 49

For the numerical check of the scheme, calculations by the "large particles" method were compared with the calculations by means of the schemes of Lux-Vandroff (Fig. 48).

It is seen from this figure, that the smearing of the shock wave is less in the "large particle" method and it fairly exactly describes the flow in the immediate vicinity of

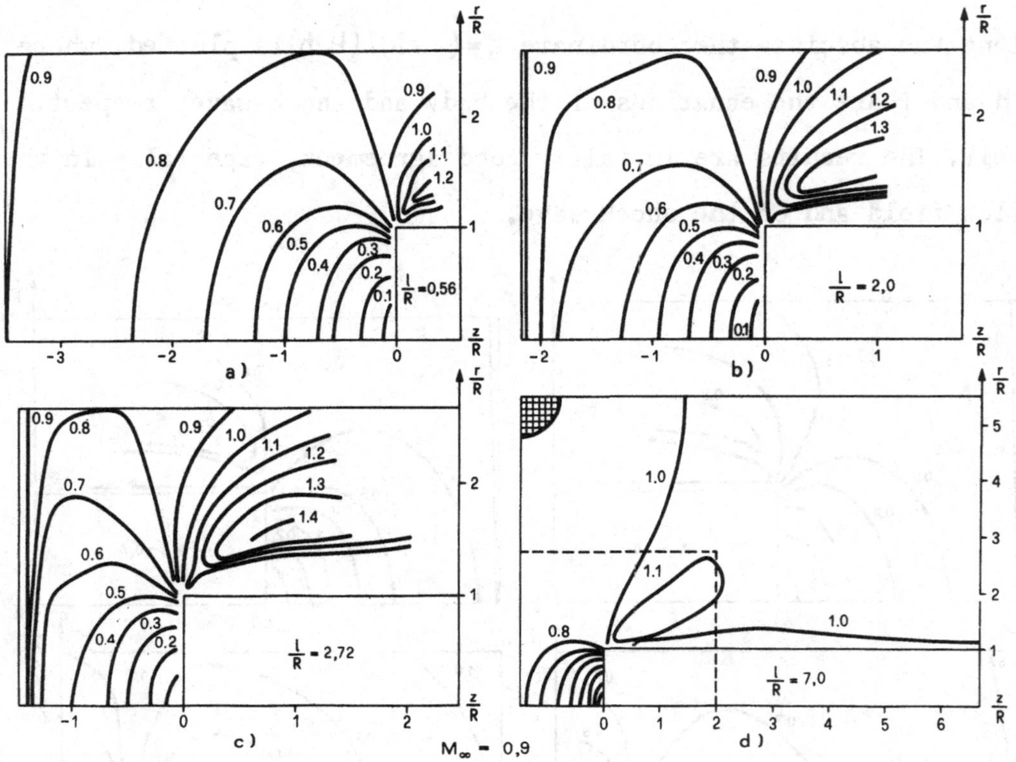


Fig. 50

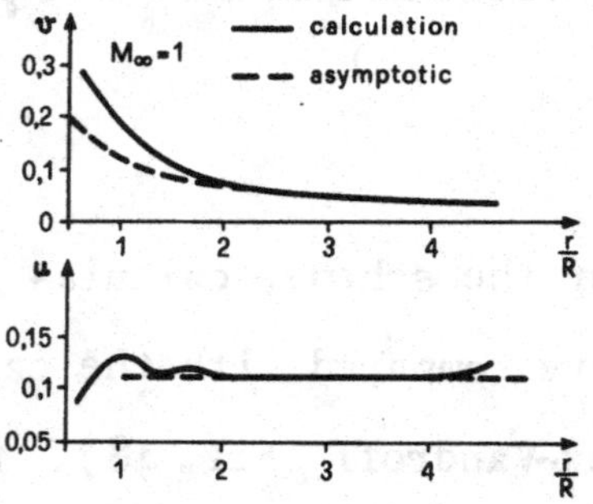


Fig. 51

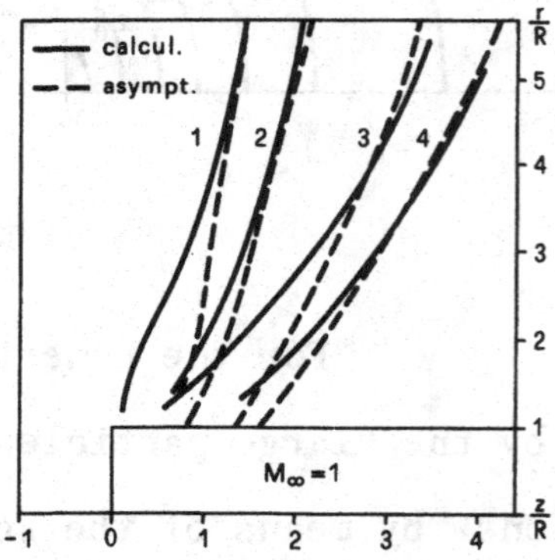


Fig. 52

the corner point. For example, the particular "large particles" method as compared to the Lux-Vandroff method should be admitted, to be more exact since local characteristics of the shock wave, of the sonic line, of the vicinity of the corner point are more apparent here.

By means of the above unsteady "large particles" method an attempt was made to construct such flows by considering the flow around an infinite axisymmetric cylindrical flat-nose body. Use was made of the above "sweeping through" method not only for calculating the transition through the sonic velocity down to purely subsonic flows.

Lines $M = \text{const}$ are constructed for the conditions of the transonic flow $0{,}6 \leq M_\infty \leq 1{,}0$ (Fig. 49) around the flat-nose body. The value of the critical Mach number is equal to $M_{\infty cr} = 0{,}7$ here. The same value of $M_{\infty cr}$ for the cylindrical flat-nose body is obtained in the experiment. Besides, in Fig. 49d the profiles of the Mach number along the axis of symmetry are given, and in Fig. 49d the thickness of the shock waves are shown (the thickness of the shock wave was determined, as it is shown in Fig. 41).

To eliminate, if possible, the influence of the boundary conditions on the flow pattern, the calculation for "overcritical" velocities were carried out with different values of the ratio of the length of the cylindrical flat-nose body $\ell$ to its radius $R$. In Fig. 50 results of such calculations for $M_\infty = 0{,}9$ are shown. The field ahead of the body was determined rather quickly, and for the cases in question it practically remained unchanged at distances up to $1{,}5\,R$ to the cut. The flow to the right of the cut is stabilized with $l/R \cong 2 \div 4$. (Boundary conditions at the left limit become important as the end is moved out further and calculation network remains fine.)

For the numerical check of the scheme (selection of the dimension of cells) the calculations of the transonic flow conditions were also carried out by means of different calculation networks. Using a more course network one may considerably enlarge the dimensions of the calculated region. In Fig. 50d lines $M = \mathrm{const}$ $(M_\infty = 0{,}9)$ are given for the calculated field which is two times larger than usual. The region calculated by the fine network is, in this case, internal with respect to the field calculated calculated by a rough network (in Fig. 50d it is shown by primes ). The comparison of both solutions indicates that they are in good agreement in the zone of the corner point, in the transonic region etc.

In Fig. 51-52 the asymptotic form of analytical singular lines (primes) is compared with the results of the numerical calculations (solid lines). Here (1) - the sonic line, (2) - the boundary characteristic, (3) - the line along which the velocity vector is horizontal, (4) - the shock wave. We see, that good agreement is observed at a distance $> 3\,R$ from the body. The difference near the surface of the cylinder is, apparently, due to the influence of viscous effects.

In conclusion, we shall give on figure illustrating the flow around the finite body obtained by the particular method. As was mentioned above, approximated viscosity is useful in producing surfaces with discontinuities separating the region of the inviscid flow. In Fig. 53 the nose shock wave ($M_\infty=2$) is seen distinctly, due to the flow with discontinuity a rarefied vortex forms near the stern of the body, a "floating" shock occurs behind the back corner point (in the figure streamlines are also plotted).(*)

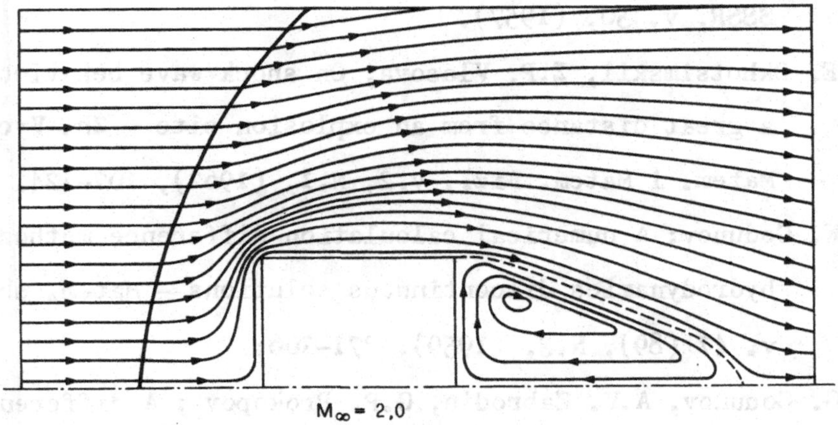


Fig. 53

(*) See also [32] , [34] , [36] - [38] .

## REFERENCES

[1] R. Courant, K.O. Friedrichs, H. Levy; Über die partiellen Differenzengleichungen der matematischen - Physik. Math. Ann., 100, 32, (1928).

[2] K.J. Babenko, G.P. Voskresenskii; A numerical calculation method of a space supersonic gas flow around a body; Zh. Vychisl. Matem. i Matem. Fiz., v.1, N.6, (1961), 1051-1060.

[3] D.E. Okhotsimskii, J.L. Kondrasheva, Z.P. Vlasova, R.K. Kazakova; Point explosion calculation with opposite pressure taken into account - Tr. Matem. Inst. AN SSSR, v. 50, (1957).

[4] D.E. Okhotsimskii, Z.P. Vlasova; On shock wave behaviour at a great distance from an explosion site - Zh. Vychisl. Matem. i Matem. Fiz., v.2, N.1, (1962), 107-124.

[5] S.K. Godunov; A numerical calculation difference method for hydrodynamics discontinuous solutions - Matem. sb., v. 47 (89), N.3, (1959), 271-306;

[6] S.G. Godunov, A.V. Zabrodin, G.P. Prokopov ; A difference scheme for two-dimensional nonstationary gas dynamics problems and flow calculation with a detached shock wave - Zh. Vychisl. Matem. i Matem. Fiz., v.1,N.6, (1961), 1020-1050.

[7] V.V. Rusanov; Calculation of nonstationary shock wave inter action with obstacles - Zh. Vychisl. Matem. i Matem.

Fiz., v.1, N.2, (1961), 267-279.

[8] A.A. Dorodnitsyn ; On one method of numerical solution of some nonlinear aerohydrodynamics problems - Tr. III Vses. Matem. Siezd., 1956, v. III, Moscow, Izd. AN SSSR, (1958), 447-453.

[9] O.M. Belotserkovskii, P.I. Chushkin ; A numerical method of integral relations - Zh. Vychisl. Matem. i Matem. Fiz., v. 2, N.5, (1962), 731-759.

[10] P.I. Chushkin ; A subsonic gas flow around ellipses and ellipsoids - Sb. "Vychisl. Matem.", N.2, Izd. AN SSSR, (1957), 20-44.

[11] P.I. Chushkin ; Subsonic gas flow calculation around an Arbitrary-shape profile and a body of revolution (a symmetrical case) - Sb. "Vychisl. Matem.", N.3, Izd. AN SSSR, (1958), 99-110.

[12] P.I. Chushkin ; Calculation of some subsonic gas flows - Prikl. Matem. i Mekhan., v.21, N.3, (1957), 353-360.

[13] O.M. Belotserkovskii ; A flow with a detached shock wave around a circular cylinder - Dokl. AN SSSR, v.113, N.3, (1957), 509-512.

[14] O.M. Belotserkovskii ; A flow with a detached shock wave around a symmetrical profile - Prikl. Matem. i Mekhan., v.22, N.2, (1958), 206-219.

[15] O.M. Belotserkovskii ; On calculation of a flow with a de-

tached shock wave around axisymmetrical bodies performed on a computer - Prikl. Matem. i Mekhan., v.24, N.3, (1960), 511-517.

[16] A.A. Dorodnitsyn ; On one method of the equation solution of a laminar boundary layer - Zh. Prikl. Mekhan. i Tekhn. Fiz., v. 1, N.3, (1960), 111-118.

[17] Yu. N. Pavlovskii ; Numerical calculation of a laminar bound ary layer in a compressible gas - Zh. Vychisl. Matem i Matem. Fiz., v.2, N.5, (1962), 884-901.

[18] O.N. Katskova, A.N. Kraiko ; Calculation of plane and axisymmetric supersonic flows with irreversible processes - Moscow, (1964), Vych. Ts. AN SSSR.

[19] O.N. Katskova, I.N. Naumova, Y.D. Shmyglevskii, N.P. Shulishnina, Experience in the calculation of plane and axisymmetric supersonic flows by the method of characteristics - Moscow, (1961), Vych. Ts. AN SSSR.

[20] O.N. Katskova, P.I. Chushkin ; On one scheme of a numerical method of characteristics - Dokl. AN SSSR, v.154, N.1, (1964), 26-29.

[21] O.M. Belotserkovskii, A. Bulekbayev, V.G. Grudnitskii ; Algorithms for schemes of the method of integral re lations applied to the calculations of mixed gas flows - Zh. Vych. Mat. i Mat., Fiz., v.6, N.6, (1966), 1064-1081.

[22] O.M. Belotserkovskii, A. Bulekbayev, M.M. Golomazov, V.G. Grudnitskii, V.K. Dushin, V.F. Ivanov, Y.P. Lunkin, F.D. Popov, G.M. Ryabinkov, T.Y. Timofeeva, A.I. Tolstikh, V.N. Fomin, F.V. Shugayev; Flow past blunt bodies in supersonic flow : theoretic al and experimental results - Edited by O.M. Belotserkovskii, Trudy Vych. Ts. AN SSSR, published by Computing Center, AN SSSR, 1966 (1st edition), 1967 (2nd edition, revised and extended).

[23] P.I. Chushkin, Blunt bodies of simple form in supersonic gas flow - Prikl. Mat. i Mech., v.24, N.5, (1960), 927-930.

[24] P.I. Chushkin ; Method of characteristics for three-dimensional supersonic flow - Trudy Vych. Ts. AN SSSR, Moscow, (1968).

[25] K.M. Magomedov, A.S. Kholodov ; On the construction of difference schemes for equations of hyperbolic type based on characteristic coordinates - Zh. Vych. Mat. i Fiz., v.9, N.2, (1969), 373-368.

[26] F.H. Harlow ; The particle-in-Cell Computing Method for Fluid Dynamics - Methods in Computational Physics, v.3, edited by Berni Alder, Sidney Fernbach, Manuel Rotenberg, Academic Press, N.Y., (1964).

[27] M. Rich ; A method for Eulerian Fluid Dynamics - Los Alamos Scientific Laboratory, New Mexico, Lab. Rep.

LAMS - 2826, (1963).

[28] O.M. Belotserkovskii, Y.M. Davidov ; The use of unsteady methods of "large particle" for problems of external aerodynamics - Vych. Ts. AN SSSR, (1970), 85 p.

[29] O.M. Belotserkovskii, E.S. Sedova, F.V. Shugaev ; Supersonic Flow Around Blunt Bodies of Revolution with a surface Discontinuity - Zh. Vychisl. Matem. i Matem. Fiz., v.6, N.5, (1966), 930-934.

[30] Aerophysical Investigations of Supersonic Flows. Edited by Dunaev Yu. A., Izd. "Nauka", Moscow - Leningrad, (1967).

[31] O.M. Belotserkovskii, E.G. Shifrin ; Transonic flows behind a detached shock wave - Zh. Vychisl. Matem. i Matem. Fiz., V.9, N.4, (1969), 908-931.

[32] *O.M. Belotserkovskii et. al.; Numerical investigation of modern problems in gas dynamics - Izd. "Nauka", Moscow, (1974).

[33] V.F. Diachenko ; On one new method of the numerical solution of nonstationary gas dynamics problems with two space variables - Zh. Vychisl. Matem. i Matem. Fiz., v.5, N.4, (1965), 680-688.

(*) The works 32 - 38 were added in the last time.

[34] O.M. Belotserkovskii, Y.M. Davidov ; A non-stationary "coarse particle" method for gas-dynamical computations - Zh. Vychisl. Matem. i Matem. Fiz., v.11, N.1, (1971), 182-207.

[35] R.A. Gentry, R.E. Martin, J. Daly ; An eulerian differencing method for insteady compressible flow problem - J. Comput. Phys., v.1, (1966),87-118.

[36] O.M. Belotserkovskii, Y.M. Davidov ; Computation of transsonic "supercritical" flows by the "coarse particle" method - Zh. Vychisl. Matem. i Matem. Fiz., v.13, N.1, (1973), 147-171.

[37] O.M. Belotserkovskii ; Numerical methods of some transonic aerodynamics problems - J. Comput. Phys., v.5, N.3, (1970), 587-611.

[38] O.M. Belotserkovskii ; Numerical experiment in gas dynamics - Lecture Notes in Physics, Springer-Verlag, N.35, (1975), 79-84.

Printed in the United States
by Bookmasters

Printed in the United States
By Bookmasters